DO I KNOW YOU?

DO I KNOW YOU?

A Faceblind Reporter's
Journey into the Science of Sight,
Memory, and Imagination

SADIE DINGFELDER

Little, Brown Spark
New York Boston London

Do I Know You? is a work of nonfiction. Some of the dialogue is based on taped conversations; the rest has been reconstructed. Some names of individuals have been changed, along with potentially identifying characteristics, and some timelines have been changed. The approaches, techniques, and remedies referred to or discussed in this book reflect the author's experiences and/or opinions only, are not intended as recommendations, and should not be construed as a substitute for medical advice.

Copyright © 2024 by Sadie Dingfelder

Hachette Book Group supports the right to free expression and the value of copyright. The purpose of copyright is to encourage writers and artists to produce the creative works that enrich our culture.

The scanning, uploading, and distribution of this book without permission is a theft of the author's intellectual property. If you would like permission to use material from the book (other than for review purposes), please contact permissions@hbgusa.com. Thank you for your support of the author's rights.

Little, Brown Spark
Hachette Book Group
1290 Avenue of the Americas, New York, NY 10104
littlebrownspark.com

First Edition: June 2024

Little, Brown Spark is an imprint of Little, Brown and Company, a division of Hachette Book Group, Inc. The Little, Brown Spark name and logo are trademarks of Hachette Book Group, Inc.

The publisher is not responsible for websites (or their content) that are not owned by the publisher.

The Hachette Speakers Bureau provides a wide range of authors for speaking events. To find out more, go to hachettespeakersbureau.com or email hachettespeakers@hbgusa.com.

Little, Brown and Company books may be purchased in bulk for business, educational, or promotional use. For information, please contact your local bookseller or the Hachette Book Group Special Markets Department at special.markets@hbgusa.com.

"Ocular Dominance Columns" reproduced from David Hunter Hubel and Torsten Nils Wiesel, "Ferrier Lecture: Functional Architecture of Macaque Monkey Visual Cortex," *Proceedings of the Royal Society B: Biological Sciences* 198, no. 1130 (1977): 1–59. On page 124, with permission from the Royal Society. "Reading Rainbow Theme Song" by Stephen Horelick, Dennis Kleinman, and Janet Weir ©1981 SCH Music, Inc. ASCAP. Lyric on page 218 reprinted with permission from the publisher.

Design by Bart Dawson

ISBN 9780316545143
LCCN 2023951524

Printing 2, 2025

LSC-C

Printed in the United States of America

To my unusual brain, without whom this book would have been unthinkable.

CONTENTS

Introduction   3

1. Grocery Store Epiphany   13
2. Three Moms and a Nazi   25
3. Missed Connections   39
4. Hacking the System   57
5. Your Brain's Rosetta Stone   69
6. Oblivious Ineptitude   75
7. Fear, Unmasked   97
8. Student Driver   113
9. Sadie Vision   131
10. Hollywood Meets Science   145
11. Video Game Therapy   161
12. We're All Making the Same Mistake   177
13. Quantifying Quirkiness   197
14. Triangulating the Truth   213
15. Gullible's Travails   233
16. Comic Incompetence Is the Brand   247

Appendix: Practical Advice   267
Acknowledgments   271
Notes   275
Index   285

DO I KNOW YOU?

INTRODUCTION

As my husband drops me off at a state park, I feel like a kindergartner on the first day of school. "Have fun!" he says, before driving away. A group of outdoorsy-looking ladies are staring at me, and I know why: I'm a forty-two-year-old woman who depends on her husband for rides. I'm here to make friends, but all I want to do is run and hide.

Instead, I introduce myself with false confidence. When the members of the hiking group respond in kind, I attempt to hitch their features to their names. Jean is wearing jeans, but she probably owns other pants. Sandy has sandy-blond hair, but so do at least five other women standing around her. A woman named Dale introduces herself. All I can think is how unfortunate it is that she doesn't have prominent front teeth or chipmunk cheeks.

I'm very disappointed in my performance, because I *studied*. Using Facebook, I snooped on all these ladies before I arrived. I made flash cards with their photos on one side and their names and fun facts on the other. I could fish these out of my backpack for a refresher, but that would be creepy.

Finally, I see someone who stands out. Kathy is a tiny seventy-something lady with long, wild, wavy hair. In her Facebook profile picture, she's playing a bamboo flute at a Renaissance festival.

"You look familiar," I say, which is not technically a lie. "Do you play flute? Or recorder?"

"No," she says, looking vaguely annoyed. (I later discover that she was just posing with that flute.)

My heart is now racing, and my face feels hot despite the early-spring chill. At the edge of the parking lot is a freestanding bathroom. Would anyone notice if I just sauntered over there and hid until they left?

I'm also tempted to fake it—to plaster my face with a dumb smile, keep conversations vague, and avoid using proper nouns—but I've promised myself that I'm not doing that anymore. I'm going to come clean about my disability.

"So, just so you know, I'm faceblind," I tell Sandy (I think?) as we hoof it up a steep hill.

"Oh, me too!" she replies. "I'm just *terrible* with names."

Inwardly, I roll my eyes and sigh. My problems go much, much deeper than forgetting names. I forget *faces*. And due to my particular combination of neurological disorders and quirks, I also forget *people*—not just their names but their very existence. And when I do manage to dig up someone's name, I get so excited I feel the need to tell everyone. My introductions, however, often backfire, because I tend to forget people's relationships to one another. ("Sadie, I know Josh. He's my brother.")

I've always known that I'm a little quirky, but for forty-odd years I didn't realize I was having trouble with tasks that other people find trivially easy. From where I sit, if you're like 98 percent of the population, you're a face-recognition virtuoso. Thanks to

INTRODUCTION

a specialized chunk of brain called the fusiform face area (FFA), you can simultaneously take in a collection of features—left eye, right eye, nose, mouth, eyebrows, cheeks, freckles, dimples, hairline, and a zillion other things—and make note of distinctive characteristics while concurrently processing their relationships to one another. Then you create a three-dimensional model of that face that allows you to recognize it from different angles and in all kinds of lighting conditions.* You're rendering graphics in your head faster and more accurately than the most advanced computer, and you aren't even trying.

So, while you may forget someone's name from time to time, I bet you have no trouble following movies, even if the main character puts on a hat, or—worst of all—gets a makeover. (Me watching *Pretty Woman*: "Who's that? What happened to the prostitute?")

I can't do any of these things, and I only recently discovered why. I have prosopagnosia, colloquially known as faceblindness. As a result, I lack some of that specialized face-recognition software, so I end up using the generic program, the one that most people use for recognizing, say, a nice rock, or a beloved pet hedgehog.

I take a breath and prepare to explain this to Sandy. But then I hesitate. Maybe Sandy and I *are* talking about the same thing. Before I knew about prosopagnosia, I, too, told people that I was bad at names. Maybe Sandy is also faceblind, or maybe she has a completely different neurological disorder that makes remembering names tough for her.

And anyway, why do I think I'm so special? Everyone has *something*, right? Scientists estimate that faceblindness affects

---

* If you're thinking, *No I can't. No one can do that,* you, my friend, may be faceblind.

about six million Americans, the vast majority of whom have never been diagnosed. But that's not even close to the most common visual or cognitive disorder. There are three hundred million colorblind people in the world, and hundreds of millions more whose brains have a quirk they don't even know about. I personally know two people who weren't diagnosed with ADHD until their thirties, and one who discovered she was dyslexic at forty, not because anyone referred her for a diagnosis but because TikTok's uncanny algorithm started serving her videos on the topic. A colleague didn't realize she lacked a sense of smell until she was *twelve*. When other people talked about smelling things, she just thought she was "doing it wrong"—and no one realized anything was amiss until she failed to notice the overwhelming stench of plastic burning on the stove next to her.

As for me, I failed to notice that I was making mistakes no one else makes—getting into strangers' cars, getting lost in my brother's three-bedroom house, making plans to meet up with someone and then being surprised by who showed up—for forty years. Uncertainty, improvisation, and wacky mishaps make up the fabric of my life, and I never thought to question it until I tried to write some of these silly stories down. That's when it finally dawned on me that something weird was going on.

So, I did some research, signed myself up for a study, and made a shocking discovery: I am not just a little bad at remembering people. I am legit faceblind. If you imagine a normal curve, you'll find me way over on the left. I'm in the bottom two percent, two standard deviations below the mean. I'm about as good at face recognition as Elon Musk is at branding.

When I got this verdict, I accepted it. My dad, not so much. "Are you sure you're really trying?" he asked. Also: "Why worry about something you can't change?"

INTRODUCTION

Soldier onward, bluff your way through, figure it out, and make it look easy. This is the Dingfelder way of life, and it's served me well. There's only one little problem: A key part of being an effective con artist is conning *yourself*. And once I began to see the cracks in my neurotypical facade, I couldn't look away.

How did I fail to notice this "disorder"—one that had been quietly shaping my entire life? What *else* don't I know about myself?

A lot, as it turns out.

In the following chapters, I will visit neuroscience labs across the country. I will spend around thirty hours in fMRI machines and dozens more taking computer-based tests, sometimes with electrodes affixed to my head. Scientists will write papers about me! And I will learn that at some point in my brain's development, it took a detour onto a path less traveled. In addition to being faceblind, I am also stereoblind. Plus, I have aphantasia (a blind mind's eye), SDAM (severely deficient autobiographical memory), and probably other things that don't have names yet.

So. Welcome to my midlife crisis. There will be no fast cars or sexy pool boys, but there will be answers to questions that have dogged me my entire life. Mysteries like: Why didn't I ever learn how to drive? Why hasn't anyone ever asked me out on a date? Why was I so lonely as a kid, and how did I manage to make so many friends as an adult? (And why, despite having so many friends, do I still feel lonely?)

What I learn about vision, memory, and imagination will captivate me—millions of miracles happening inside our brains every minute! What I discover about the places where my brain falls short of miraculous will force me to reinterpret major events from my past and mourn losses that I didn't even know I'd suffered. I'll upset my parents with prying questions, worry my husband with excessive crying, and feel sorry for the kid version of myself, who often felt misunderstood and who still feels that way sometimes.

As I've struggled to see myself more clearly, I've also discovered something about everyone else: There's a stunning amount of hidden neurodiversity in the world. Your best friend, your spouse, your boss—their conscious experience might be completely different from yours, and neither of you may have any idea!

Remember the viral photo of that dress that looked white and gold to some people but black and blue to others? We were all looking at the same image and seeing completely different things. This isn't a fluke—this kind of thing happens *all the time*. The world is full of ambiguous information, and different brains come to different conclusions.

Then there are people's inner lives. The variety of ways people experience being awake and alive is, frankly, mind-boggling. If you don't believe me, start asking your friends questions like:

> Do you have an inner monologue?
> Do you hear it in your mind's ear? Is it in your own voice?
>> Is it like eavesdropping on your own thoughts, or is it more like a commentary on what you're doing?
>
> (Me? Most of the time my mind is quiet.)
>
> When you're reading a novel, do you "see" the characters in your mind's eye?
> Do you visualize settings and enjoy lush descriptions?
> (Lucky you! I see nothing but words on a page, and I often skim descriptive passages to get to the plot.)

INTRODUCTION

> Can you revisit important moments from your past in vivid visual and emotional detail?
> Are these memories in color or black and white?
> Do you experience them in the first person or third?
> Are they moving or still?
> (All I remember of my past are the stories I tell about myself—all words, no pictures, and muted emotions.)

You may find my experiences difficult to believe. I find yours equally unlikely. I mean, if other people can hallucinate at will, surely I would have heard about it...right?

Actually, there have been clues: My yoga teacher, for instance, who's always asking me to do impossible things, like "Imagine your hip bones are headlights." Many people can follow this instruction! For the longest time, I thought the opening credits of *Reading Rainbow* were trying to trick kids into thinking that reading is like watching cartoons. Evidently, though, neurotypical people visualize whatever they are reading about. (No wonder you're all so slow.)

"Counting sheep"? For me, it's exactly the same as regular counting, except...fluffier? I guess I didn't think about it that much.

What's clearly a metaphor to one person can be quite real to another.

Then there's the issue of credibility. When I try to describe my conscious experiences, why should you believe me? Why should I believe *myself*? After all, I'm clearly a master of self-deception, having thought, for decades, that I was basically neurotypical.

These seemingly intractable problems are why, in the 1950s, the field of psychology stepped away from studying inner experiences.

Psychologists wanted to be taken seriously as scientists, and scientists study things that can be observed and quantified. Questions like "Is the red I see the same red you see?" seemed best left to philosophers and potheads, while less interesting, more concrete issues like "How can I make this pigeon work harder?" were taken up with great rigor.

Fortunately, the tide is turning, and scientists are beginning to study internal experiences again. The advent of fMRI in the '90s kicked things off by capturing snapshots of the mind at work. Since then, psychologists and neuroscientists have come up with lots of other clever ways to corroborate or contradict what we claim is going on inside our skulls.

For instance, you can ask vivid visualizers to imagine a brightly lit shape and see whether their pupils contract. (They do![1]) If you put them into an fMRI machine and assign them the same task, you can see how much their occipital (vision) cortexes activate. (A lot![2]) Make them read a violent passage in a book and see if they break a sweat. (They get positively damp compared to people like me.[3])

As scientists begin taking inner experience more seriously, they are discovering entire continents of wild, beautiful, and largely uncharted neurodiversity. And yet, even with this information in hand, it's difficult to break out of the assumption that everyone's conscious experience is pretty much like your own.

To explain why, I offer a parable:

An old fish passes by a school of youngsters and says,
    "Hey, boys, how's the water?"
"What's water?" the little fish reply.

INTRODUCTION

For these young fish to understand what water is, they need something to compare it to—maybe air, or the vacuum of space, or something *really* weird, like the experience of swimming through melted chocolate. Well, dear reader, I'm the fish who just figured out that I live in a sea of melted chocolate, and I'm going to attempt to describe what it's like so that you can see whatever substrate it is *you're* swimming through.

The best fish to ask about chocolate vs. water is, of course, one that's swum in both—and that's part of the reason why, in the coming chapters, I'll be trying to experience the world more like neurotypical people do. I'll attempt to teach my brain to quickly assess distances between facial features. I'll try to learn to see in 3D by playing virtual reality video games that are not yet FDA approved. I'll work on my powers of visualization with educators who teach kids to use their mind's eye to spell or do mental math. If that doesn't work, I'll immerse myself in a sensory deprivation tank—or try shrooms, maybe? (I'm leaning toward no on that one.)

"But, Sadie," you say, "how can you claim to celebrate neurodiversity while attempting to quash it in yourself?" Good question. I'm glad you asked. There are advantages and disadvantages to all of my cognitive and perceptual differences. Take stereoblindness, for example: When your eyes don't work together, it's hard to catch a ball, walk on uneven ground, or merge your car onto a highway. On the other hand, seeing out of only one eye at a time can give you a slight edge as an artist. There's evidence that many famous artists—Gustav Klimt, Edward Hopper, Andrew Wyeth, Marc Chagall, Frank Stella, and Man Ray, to name a few—had misaligned eyes.

Even though I know this, I'm still going to try to learn to see in 3D. I've had the entire first half of my life to parlay my stereoblindness into artistic ability, and all I can draw are cartoon mermaids.

This could all be a huge mistake, of course. But I take some solace in the fact that I've been living with my weird brain for more than forty years. If I manage to learn to visualize, for instance, it probably won't unleash a flood of unwanted imagery upon my poor tortured soul. I'll be lucky if I can figure out how to conjure the occasional dim and fleeting beach sunrise. If I were younger, I'd be more concerned.

I do want to succeed, though. I want to at least glimpse how the other 98 percent live. What is it like to know, with total certainty, who someone is? What does a tree look like in 3D? Isn't it distracting to "see" things that aren't actually there? Is it really possible to think before you speak?

Of course, I may end up regretting my choices or even changing my mind.

So, let's get going!

"Let's get going" is also more or less what Sandy says to me when she notices the two of us falling behind the other hikers. The pandemic has atrophied my social skills, and I've just wrapped up a monologue very much like the one you just read, except with more panting. I'm out of shape and out of breath, but what's really slowing me down is my stereoblindness. Steep, uneven paths are tough to navigate without depth perception.

I know exactly what this situation calls for: a lecture on stereovision!

Just kidding. I want these women to think I'm at least a little normal.

"Don't wait for me," I say, dropping to all fours. "I'll get there eventually."

1

GROCERY STORE EPIPHANY

I'll never forget that look of disgust in my husband's eyes—disgust mixed with disappointment—the time he caught me with a bag of frozen Lender's Bagels.

I should have the upper hand when it comes to bagel snobbery, as I'm the Jew in this relationship. But as Steve's expression made plain, he has strong feelings about frozen bread products. I tried backpedaling. "I used to eat these as a kid," I said, quickly returning them to the Safeway freezer. "I wasn't thinking of *buying* them," I added. "I just wanted to show you." Steve seemed unconvinced.

This is why, a few months later, I'm feeling positively triumphant when I spot Steve grabbing a jar of Safeway's Signature Select peanut butter. Just the other day, he'd been extolling the superiority of homemade nut butter. What a hypocrite!

"Since when do *you* buy generic?" I say, plucking the jar out of our cart and then brandishing it like it's exhibit A in a murder trial.

The look on Steve's face this time isn't disgust. It's...fear? I'm about to turn around to see what has him spooked, when something really weird happens. His features start looking sort of...wobbly. Unstable. As if they're affixed to his head by little invisible springs.

All at once, I see that this man is not my husband. He's a Steve-shaped stranger who's wearing almost the exact same coat. What a nasty trick!

I drop the jar and flee, without another word.

When I find Steve in the checkout line, I'm flushed with embarrassment. But I'm already working on turning this experience into a party anecdote.

"I bet he's thinking, *Wow, market research has gotten really aggressive,*" I say.

Steve laughs, and I'm starting to feel better. But on the drive home, a disturbing thought bubbles up from my subconscious. *Other people do not make this kind of mistake.*

A few years later, I'm in the Baltimore Museum of Art, reporting on a new exhibit. I watch as a young woman browses the gallery with detached interest until she comes upon a group of photos that puncture her composure. "Oh," she gasps, leaning in to read the title of the piece: *12 Assholes and a Dirty Foot*. It delivers exactly what it promises.

A few minutes later, the artist materializes in the gallery.

"Is that him?" someone half whispers.

Of course it is. With his pencil mustache and Hollywood smile, the cult filmmaker John Waters is unmistakable. For me, his arrival isn't a surprise, but it is a change of plans. I was originally supposed to interview him after museum hours, but scheduling issues conspired to create this interesting situation.

The museum PR lady introduces us to each other, and Waters begins showing me around. We stop beside a large line drawing of...a tick?

"No, it's crab lice," Waters says. "Now, that's a species that's going extinct. Because, you know, young people don't have pubic hair anymore."

Pubes gone, no crabs, I scrawl on my notepad.

I look up and see that we are being trailed by a growing crowd of Waters's fans. A born showman, the filmmaker projects his voice, subtly acknowledging the eavesdroppers, which emboldens more to join the group. Pretty soon, we're encircled by more than a dozen women, ages thirty to fifty. ModCloth dresses and chunky necklaces abound. I fit right in, and unfortunately, so does the museum PR lady. I scan the scrum of Ms. Frizzles, looking for a telltale sign, a museum lanyard, a clipboard—something!

A woman with a large purse and an "on duty" vibe meets my increasingly frantic gaze. She must be my gal.

"Is there a place with natural light where we could take some pictures?" I ask.

"I don't know," she says with a shrug.

Waters is now looking at me like *I'm* the weirdo.

"Do you know where the museum lady went?" I ask. He spots her by the entrance—picking her out of the crowd as easily as if she were a neon sign.

"Sorry, I'm a little faceblind," I say with an awkward laugh.

Later, while transcribing my interview, I listen to a replay of the whole incident. There's me, sounding confused. There's Waters, locating the museum rep like it's no big deal. There's me, making a lame joke. But it doesn't sound like a joke. It sounds... true.

I first learned about faceblindness in 2010, when I happened upon an article by Oliver Sacks in *The New Yorker*. Like me, Sacks had long assumed that his difficulties remembering people were the result of general absent-mindedness. It wasn't until halfway through his life that Sacks — a neurologist! — realized that he had a neurological disorder (or if you prefer, difference).

I read with astonishment as he described moments from his life that could have been lifted from mine. Like me, Sacks once accidentally snubbed his therapist in public, and she refused to believe it was a simple oversight. Like me, Sacks once thought his own reflection was a stranger with a poor sense of personal space.

Despite these similarities, I was hesitant to diagnose myself with Sacks's disorder. For one thing, it seemed presumptuous to think that I had the same condition as this world-famous author. And while we've had some of the same experiences, we are very different people. As he put it, "I think that a significant part of what is variously called my 'shyness,' my 'reclusiveness,' my 'social ineptitude,' my 'eccentricity,' even my 'Asperger's syndrome,' is a consequence and a misinterpretation of my difficulty recognizing faces."[1]

This didn't sound like me at all. I am not the least bit shy — in fact, I talk to strangers every day. Sometimes it feels like all I do is talk to strangers.

I tossed that *New Yorker* into my recycling bin, and I didn't think about faceblindness again until the fall of 2018.

What happened? A few things. I'd been trying to write down some of my funny anecdotes for posterity, or perhaps even publication, but they kept coming out weird. The grocery store story, for example, was simply not adding up. As a reporter, I've developed an instinct for knowing when someone isn't telling me the whole truth, and I couldn't shake the feeling that I was lying—to myself!

Then, while transcribing my John Waters interview, I hear myself saying that I'm faceblind, and it feels like the lead I've been looking for.

I abandon my assignment and go to PsycNet—a research database from my old job at the American Psychological Association that I should not, technically, still have access to.* I type "prosopagnosia"—the technical term for faceblindness—into the search bar, and the database comes up with more than a thousand records, going all the way back to the 1940s! There are studies, dissertations, book chapters, case reports—a metric ton of material on what I assumed would be a niche topic.

Why, I wonder, *is there so much research on this rare disorder?*

Three men are staring at me, their expressions unnervingly blank. On my laptop screen, each face displays as roughly two by three inches, photographed from the shoulders up, like passport photos. They are set against the same black background, and their hair has been cropped off. They're not identical, but their features all blend together into a single, vaguely menacing-looking white guy.

I'm trying to pick out a particular face—one that I just stared at for twenty seconds. For me, this is like trying to identify a particular penguin on an Antarctic march. I am failing the Cambridge Face Memory Test, and I hate failing.

---

* When we made outgoing calls, our number came up as "American Psycho."

"What are you groaning about?" Steve asks. He's lying beside me in bed, solving chess puzzles—which is his idea of unwinding after work.

As with many computer nerds, I think it's fair to say that Steve's more interested in ideas than people. Despite living in the same building for fifteen years, he only knows the names of maybe five of his neighbors. He also has a total of three or four friends, a tiny crew, especially when compared to my cast of hundreds.

To feel better about myself, I make Steve take the face memory test while I watch over his shoulder.

He flies through it, clicking on the stone-faced guys as if they are old pals.

"How are you doing that?" I ask, astonished and a little angry.

He shrugs, blithely unaware of my distress. "I don't know, I'm just"—a brief wave of his hand—"doing it."

In less than twenty minutes, Steve breezes through a seventy-two-item test that took me nearly twice as long. This seems like magic to me, but he thinks what he's doing is perfectly normal. As I would later learn, we are both right.

Want to give it a try? Here's an item from a different section of the test.

You've got twenty seconds to memorize this guy:

GROCERY STORE EPIPHANY

Do any of these faces look familiar? You have three seconds.

How about these?

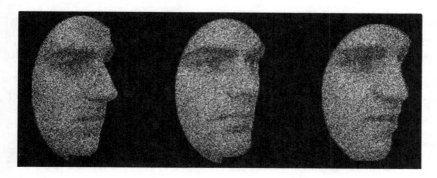

The answers are at the bottom of the page.* How'd you do? For the first question, 79 percent of neurotypical folks, 47 percent of faceblind people, and 100 percent of super recognizers get it correct. (More on them later.)

The second group of faces adds visual "noise" to the photos. For this kind of question, 68 percent of neurotypicals, 38 percent of faceblind people, and 97 percent of super recognizers get it correct.

Steve's score for the entire test, 80 percent, puts him right in the middle. From all the studies I've been reading, I know that

---

* First question, the face on the far left. Second question, the face on the far right.

face recognition abilities are highly heritable, so I send the test to my brother. If I'm terrible at this, Saul should be too — but he lands on the *high* end of normal, with 89 percent correct.

Me? I score 58 percent.

I expected to come in a little below average. I did not expect to score in the same range as people who have been literally shot through the head.

When I send my score to Joe DeGutis, the Harvard neuroscientist who had sent me the test, I include a note of apology.

"I think I could have tried harder, or maybe I was tired or something," I write.

DeGutis is one of several scientists I've been emailing with questions that I hope sound casual but professional: How do people typically recognize each other? Is it normal to forget what people look like two seconds after meeting them? Is it weird that I was faster to notice the wrong label on a peanut-butter jar than the wrong face on my husband?

I tell the scientists that I'm a journalist (which is true) and that this is for a story (which is not). What I'm really asking is: Should I be scared? Is there something wrong with me?

DeGutis's research assistant, Alice Lee, sends me some more tests, including a Famous Faces Test, which I find to be more reasonable. Brad Pitt, check. Hillary Clinton, obviously. Isabella Rossellini — yes! I have seen her in person and even interviewed her once. Rossellini again, but older. You can't trick me, wily test designers.

"I think I did pretty well on the celebrity test," I write Lee.

Boy, am I wrong. The photos I thought were of young and old Isabella Rossellini turn out to be Scarlett Johansson and Margaret Thatcher. (Sorry, Isabella!)

"The thing about movie stars," I say to Steve, "is that they all look basically the same."

It's 1 a.m. and I've started downloading more papers on prosopagnosia. There's so much to learn; I just can't help myself.

Steve rolls his eyes and closes my laptop. "Go to sleep," he says.

I work at the *Washington Post Express,* the *Washington Post*'s sassy little sister. While my colleagues are out there defending democracy, I'm reviewing all the public toilets on the National Mall. We haven't been selling a whole lot of ads lately, and many of us believe that the only reason we haven't been shut down is that the mothership has forgotten we exist.

I don't want to ruin this situation for everyone, so when I go to visit my friend David Rowell, on the seventh floor, I do my best to fit in. Rowell is an editor for the *Washington Post Magazine,* and I can't remember where he sits, so I stride confidently in a random direction. My plan is to methodically search the entire floor—a strategy that, incidentally, will also provide me with a thorough overview of the communal food situation.

Today, the pickings are slim—just one small platter of cookies that was left out overnight. Given recent rumors of rats, they are probably not fit to eat...

"Hey, Sadie." I hear a scratchy voice with a slight Southern drawl—such a distinctive sound. I look up and shoot Rowell a closed-lipped smile, as my mouth is full.

Not long after Jeff Bezos bought the *Post,* he moved us from a musty old building to a bright modern office with lots of glass walls. Then came the software developers—young people, casually attired, with flashy sneakers and candy-colored hair. In his rumpled dress shirt and tie, Rowell looks like an extra who wandered off the set of *All the President's Men* and into an Apple Store.

We find an empty conference room and I pitch my story.

"I've been invited to join a Harvard study to find out if I have a rare neurological disorder known as faceblindness," I say.

DeGutis and his colleagues are testing a computer-based training program that's meant to help people learn to process faces more efficiently, I explain. My plan is to go to Boston, take a lot of tests, and then complete the sixteen-week training program.

"Do you think you're actually faceblind?" Rowell asks.

"Nah," I reply. "But I'm definitely below average."

I explain to Rowell that I probably just flubbed the test because I was tired. But since my poor showing qualifies me to be in a study, it seems like a good excuse to write about a fascinating disorder. I just hope that everyone isn't too disappointed when it turns out that I don't have it.

Rowell gives me the green light and says the magazine will pay for my plane ticket to Boston. I forget to ask about a hotel, but that's okay — I have a lot of friends who live in the area.

My first thought is that I should stay with one of the Anns, who were my best friends in college. Creatively named after their respective hair colors, Red Ann and Brown Anne lived in the same dorm, majored in education, and married their respective college boyfriends. Now they both work in Boston-area public schools and have a number (three? four?) of kids.

As I start to email the Anns, it occurs to me that I haven't been in touch with either of them in years. It feels impolite to ask for a favor out of the blue. Maybe I should stay with my friend Pam instead. We weren't super close in college, but I've done a better job of keeping in touch.

GROCERY STORE EPIPHANY

"I got into Harvard by failing some tests! Now they want me to come to Boston to take some more. Can I crash on your couch?"

"Sure. Give me a call tonight," she writes back.

I'm starting to get excited about this story. I feel like I'm in line for a mildly thrilling carnival ride—but what I think is a Ferris wheel is actually a rocket bound for outer space.

THREE MOMS AND A NAZI

On October 22, 1944, a volley of Allied artillery fire interrupted a military briefing, making matchsticks out of a German army post on the East Prussian front. Everyone present died instantly—everyone except a thirty-six-year-old lieutenant we'll come to know as H. A.

After regaining consciousness, Lieutenant H. A. dragged himself back inside the smoldering building, saw his mangled comrades, and passed out again, hitting his head on the floor. The next time the lieutenant woke up, he was bouncing along in a German field ambulance. His head collided with a steel bar on the ambulance's ceiling, knocking him out for a third time. (We don't know what happened, but I bet it involved a medic who failed to earn his Hitler Youth gurney-securing badge.)

Thoroughly battered and thrice concussed, the lieutenant eventually made it to a hospital, where he met a young psychiatrist named Joachim Bodamer.*

In a paper written after the war,[1] Bodamer described Lieutenant H. A.'s injuries and resulting deficits: The left side of his body was weak, and his left leg was completely paralyzed. He'd lost half his field of vision in both eyes, and he was newly colorblind. Oddly, he had spells during which it seemed like time was on fast-forward. But the lieutenant's most intriguing symptom, as far as his psychiatrist was concerned, was that he couldn't tell people apart.

"He did not recognize people coming to see him at first, only once they had spoken. He had noticed that his hearing had improved remarkably, and he could always identify passers-by by their footsteps, and claimed he never erred. He always knew who was coming, so his disorder did not strike anyone, hardly even himself, for a long time."

To test his theory, Bodamer showed Lieutenant H. A. a stack of photos of famous people and statues, and the lieutenant could

---

* When I read about Bodamer, I wondered if he had been involved in the Nazi program of killing mentally ill and disabled people. Ingo Kennerknecht, a professor emeritus at the University of Münster, has also been trying to answer this question, and he hasn't found anything conclusive. However, circumstantial evidence doesn't look good for Bodamer. In 1933, the Nazi government passed the Law for the Prevention of Offspring with Hereditary Diseases—a law (modeled after US legislation) that allowed for the forced sterilization of people with mental illness, learning disabilities, physical deformity, epilepsy, blindness, deafness, or chronic alcoholism. Some 400,000 people were operated on under this program and upwards of 5,000 died. This was just the preamble to the Nazis' flat-out murder of 200,000 people with disabilities, which began in 1939, two years after Bodamer joined the staff at the state psychiatric hospital in Winnenden. These horrific programs required the extensive and enthusiastic participation of many doctors and nurses. "Bodamer, as a psychiatrist, must have been involved," Kennerknecht says.

only recognize one: Hitler. What gave him away? "The mustache and the parting," the patient explained.

Bodamer then convinced Lieutenant H. A.'s wife to don a nurse's uniform and asked him to pick her out of a lineup of nurses. The lieutenant looked from nurse to nurse and—after a great deal of prodding—hazarded a guess. Much to his relief, he got it right. "Something in her expression looked familiar to me," he explained.

Earlier in the war, Bodamer had come across a similar case, a twenty-four-year-old infantryman who was shot clear through the head with a large-caliber bullet, who Bodamer called Uffz. S. Uffz. S. lost his ability to see for two weeks, but by the time he met Bodamer, the soldier could see fairly normally. His main complaint was that he could no longer recognize faces—including the face of his own mother, whom he passed in a train station without realizing it.

Russia's army was closing in from the east. Allied forces were pressing in from the west. The air was thick with the smoke of burning cities. But Bodamer couldn't stop thinking about these two patients. If we can lose the ability to recognize something as specific as a face, does that mean the brain has little modules devoted to every kind of thing we recognize? Beneath our awareness, are our brains actively pulling the world apart and putting it back together? If so, what we take as objective reality is alarmingly flimsy and fragmented.

Was Plato right? Are our perceptions merely shadows on the wall of a cave?

This strange syndrome, Bodamer wrote, could help scientists unlock mysteries of human perception, memory, and even consciousness. But what to call it? The psychiatrist proposed a combination of the Greek words for "face" (*prosop*) and "not knowing" (*agnosia*): prosopagnosia.

Like many men of his generation, my ninety-year-old grandpa is obsessed with World War II, so I'm excited to tell him about my recent dive into Nazi medical history. But when we arrive at his house the day after Thanksgiving, my four-month-old nephew, Ari, steals the show.

I can't possibly compete. Ari is adorable—a blue-eyed cherub with wispy blond hair. Everyone wants to hold him, but the moment he's out of his mother's arms, he starts making little sputtering noises, a warning before the full-out wail. Grandma turns Ari around so that he can see his mom, Katharine, and he visibly relaxes.

"He looks just like his mommy," my grandma coos. I murmur in agreement, but to be honest, I do not see the resemblance. Katharine is a long-limbed, hazel-eyed beauty—a dead ringer for Blake Lively. Ari is a chubby baby, and babies all look alike. That's why hospitals give them barcodes.

Grandpa is kvelling over Ari's grip strength when the back door creaks open. An unexpected guest walks in—and it's my mom!

"Mom!" I exclaim. "What are you doing here?"

My mother lives in Colorado, and she wasn't planning on coming to Thanksgiving in Florida this year. At least, that's what I thought. Maybe she changed her mind? I go to hug her and then stop short. Something seems off. Her shirt? No, it's her boobs. They are *way* too big.

And yet, this lady looks *exactly* like my mom. Same long blond hair, same oval face, same perky energy...

"Hey, Sadie," she says. Something clicks into place. This buxom lady isn't my mom; she's my mom's sister!

"Aunt Karen—you dyed your hair," I say.

I'm laughing, but my mind is reeling. I brought a pile of research to Florida with me (beach reading!), and just this morning I read a bunch of studies that came to the same conclusion: your mom is one person you should always be able to recognize.

In one of the first studies on the topic,[2] researchers brought three-day-old infants up to a screen with two windows cut into it. Peeking silently out of the windows were the baby's mom and a similar-looking stranger—both of whom had been drenched with air freshener.

The air-freshener procedure was necessary because infants have an excellent sense of smell, and they can easily sniff out their moms. Seeing, on the other hand, is not their forte. Newborns are quite nearsighted, they don't know how to separate objects from backgrounds yet, and they probably experience synesthesia. If you'd like to see the world through the eyes of a newborn, look through a kaleidoscope that's been smeared with butter and then take some LSD.

Amazingly, 75 percent of the infants recognized their moms—we know, because they stared at them for significantly longer than they stared at the stranger. (The remaining babies split their attention more or less evenly.)

Ari begins to fuss. He spots his mom in a room full of relatives and reaches out for her. Why am I feeling jealous? Oh, right, it's because I'm nearly forty, and I'm being shown up by a four-month-old.

This is alarming, to say the least.

As Bodamer's discovery percolated through the larger scientific community, it was met with curiosity and disbelief. Critics pointed out that his two prosopagnosia patients had not *specifically* lost the ability to identify faces. Lieutenant H. A. also had trouble recognizing cars, and Uffz. S. couldn't identify common articles of clothing. Perhaps they had run-of-the-mill visual agnosia, a relatively common consequence of brain damage. (Characterized by difficulty recognizing a variety of objects, visual agnosia was the chief complaint of the titular character in Oliver Sacks's *The Man Who Mistook His Wife for a Hat*.)

Bodamer argued that his patients' other visual-recognition problems were minor when compared to their profound face-recognition deficits—a pattern that pointed to a specific face-processing module in the brain. But everyone agreed that a patient with *pure* prosopagnosia would be more convincing.

For decades, the scientific community came up empty. There were a few promising cases in the '70s and '80s, but they turned out to be duds. There was a woman who lost the ability to identify faces and birds (as a fellow bird-watcher, I am not sure which is worse), and a faceblind farmer who also lost the ability to recognize his cows. It wasn't until 1993 that researchers at the National Hospital, London, found a case of pure prosopagnosia[3]—a man they dubbed "W. J.," who became faceblind after a stroke.

W. J. was deeply embarrassed by his condition. He was so worried about accidentally snubbing people he knew, he retired to the countryside, to the solitary life of a shepherd. (Wouldn't this be a great setup for a rom-com? I'd cast Hugh Grant.) After less than a year, W. J. realized that while he still couldn't recognize his human neighbors, he had no such trouble with his sheep. He could identify them by their faces alone.

W. J.'s skeptical doctors challenged him to learn the names of a set of unfamiliar sheep, and W. J. aced the test. Human faces, alas, continued to all look alike.

The shepherd's ability to identify animal faces and not human ones led the scientists to hypothesize that human-face perception takes place in a specialized module in the brain—and if that module gets damaged, it's impossible for other parts of the brain to take over.

Around the same time, a team of Toronto scientists discovered C. K.—a man who lost the ability to identify all kinds of objects *except* for faces.[4] After his car accident, C. K. thought a tennis racquet was a fencing mask, and he mistook a dart for a feather duster. But when scientists handed C. K. a photo of Boris Yeltsin or Imelda Marcos, he named them without hesitation.

Evidence was slowly accumulating for the existence of a specific face-recognition area in the brain, but prosopagnosiacs were few and far between. The disorder was in danger of becoming a medical curiosity, its potential for cracking open the mystery of vision and memory unfulfilled.

Then the internet arrived, and it changed everything.

Edna Choisser loved her son, Bill, but she didn't understand him. It was bad enough that he looked like a dang hippie, with that long, wavy hair and big, bushy beard of his. It was even worse that he'd come out as gay. But when, in the early '70s, Bill passed right by her without saying hello, she'd had enough. Was he ignoring her in public now? Like *she* was the embarrassment?

Bill Choisser later described the encounter on his website.

> We walked towards each other, and passed within two feet of each other, on a not-too-busy sidewalk in a neighborhood shopping district. The only way I know about this is because she told me about it later that night. She was not

amused at all by this incident, and she has never forgiven me for it.[5]

Over the next couple of decades, Choisser went from doctor to doctor, and they all said that he was perfectly healthy, and that his vision was fine. More than once, he was referred to a psychologist, to work on his mommy issues. Finally, in 1996, Choisser posted a message on Usenet, an early internet bulletin board.

> When I look into someone's face... I feel like the information is being routed to a black hole and discarded... Anybody on here have [this] facial recognition problem, or do you know of anyone who does?[6]

A trickle of responses soon turned into a flood, and the group began trading tips for common problems, like following the plots of TV shows (closed captions can help) and finding your kids at school pickups (wait for them to find you). Just discovering other people with the same issues made them all feel a little less crazy. As a woman named Tina put it: "Who has the time to research anything? I sure don't, so anything I read about on this BBS is helpful."

Choisser—a San Francisco–based engineer—did have time for research. He and his collaborators hit university libraries and discovered that their disorder already had a name: prosopagnosia.

Bill was not a fan of this Greek tongue twister of a word.

> Children do get this condition and it needs to be explainable to them. They also need to be able to explain it to playmates, or their teachers need to be able to. The

medical term will not play on a playground. "Face blind" will.[7]

The coinage stuck, and the online community grew.

In 1999, just a few years after Choisser's Usenet post, Pam and Dick Duchaine slipped into the back of a small lecture hall at UC Santa Barbara. As students drifted into the Saturday morning class, Pam and Dick tried to blend in. But their resemblance to the class's instructor was too obvious to go unnoticed.

"Oh, you must be Brad's parents," one woman said.

"How'd you know?" Dick asked.

"You look just like him," she replied.

When Brad arrived, he noticed that many of his students were missing, and the rest looked hungover.

"My parents were able to make it all the way from Wisconsin," he teased. "You guys barely made it out of bed."

He was one to talk. As an undergraduate at Arizona State, Brad did so much partying, he almost flunked out. Pam and Dick were relieved to see their brilliant, underachieving son finally hit his stride. He'd just passed his comprehensive exams with flying colors. Now all he had to do was pick a dissertation topic.

This is a high-stakes moment in any grad student's career. If you choose a question that's too broad or too difficult, you're signing yourself up for frustration and extra time in grad school purgatory. If it's too narrow, your dissertation committee won't be impressed enough to hand you a diploma. But if you choose well, you'll spend the next few years tilling a fertile field that might yield the rarest of fruits: results that satisfy your dissertation committee, provide fodder for publications, and capture the imagination of future employers.

But first, Duchaine had a class to teach. He began that day's lecture with a story about a woman who, while riding a bus, watched a man board with a little poodle. The next time the woman looked up, the man had a face that was identical to his dog's. Increasingly upset, the woman looked around and saw that the other passengers all had poodle faces as well. Getting off the bus didn't resolve the situation. For thirty minutes, every person this poor woman saw appeared to be wearing the face of that one little dog.

The woman had recently had a stroke in her right posterior cerebral artery, and thankfully, the poodle thing only happened once. The stroke, however, did have one lasting effect: she was now unable to tell people apart.

That night, all three Duchaines ate dinner at a seaside restaurant. While peeling shrimp, Pam bragged about her son to her friend Mary Beth, who was visiting from LA. When Pam started talking about prosopagnosia, Mary Beth's eyes lit up.

"I'm familiar with that," she said. "Matter of fact, we have a friend whose teenage son has that condition."

"Did he have a stroke or some other kind of brain injury?" Brad Duchaine asked.

"No, I don't think so," said Mary Beth. "Want me to get you in touch with him?"

This conversation left Duchaine buzzing with excitement. If this kid really had prosopagnosia *without* any sort of brain injury, it would be the second such case ever found.[*]

---

[*] The first was in 1976: H. R. McConachie, "Developmental Prosopagnosia. A Single Case Report," *Cortex* 12, no. 1 (1976): 76–82, https://doi.org/10.1016/s0010-9452(76)80033-0.

Duchaine visited Mary Beth's friends later that week, and he was astonished to find that the teenager did, in fact, have prosopagnosia. But even more surprising was the fact that the kid was in contact with dozens of people like himself—folks who had been terrible at identifying faces for their entire lives, no brain damage necessary. He'd found them through an online community of *developmental* prosopagnosiacs—Bill Choisser's website.

Duchaine felt like he'd dropped a pickaxe and hit oil—dozens of faceblind folks, all potential research subjects! This was more than enough fodder for a dissertation; it could launch his entire career.

As Duchaine studied Choisser, a similar story was playing out across the Atlantic. In 2001, German geneticist Ingo Kennerknecht received a call from a woman who had diagnosed herself after finding a few other faceblind people on German websites. Kennerknecht wasn't aware of Duchaine's research, but he also had a hunch that developmental prosopagnosia might be a lot more common than the acquired kind.

"Within half a year, we had more cases in Münster than the whole world had ever seen before," Kennerknecht told me.

For a 2006 study,[8] Kennerknecht and his colleagues polled 689 college and medical students about their face-recognition abilities. They found 17 developmental prosopagnosiacs (DPs)—a prevalence rate of 2.5 percent.

The fact that Kennerknecht and Duchaine discovered developmental prosopagnosia just a few years apart was no coincidence. In the late '90s and early aughts, the internet was just beginning to take off, and people with rare disorders were the web's early power users. They found one another and reached out to researchers—and that's still the case today.

"We don't have enough person-power to test all the interesting people we're hearing from," Duchaine later told me. "It's a wonderful time to be a neuropsychologist."

Duchaine is now looking for people with an *extremely* rare disorder known as hemi-prosopometamorphopsia. Hemi-PMO results from damage to the bundle of fibers that connects the brain's two hemispheres. People with the disorder see faces as melting, swelling, or sagging—but only on one side.

"No matter how the face is turned, no matter how far away it is and even if it is upside down, they perceive distortions on the same parts of the face," Duchaine said.

This disorder suggests that the brain separates faces from the surrounding scenery and then fits them to a standardized template, Duchaine added. The process makes it easier to match the face you're currently looking at with your stored memories—which happens to be how many computer face-recognition systems work. Less like computers, however, is the brain's apparent technique of splitting faces in half and sending one half to each hemisphere for processing. Then the brain brings the two halves back together, probably in the right hemisphere's fusiform face area (FFA)—an olive-sized chunk of brain matter, just above your ear, that's specialized for face processing. That's Duchaine's current working theory, anyway.

"That seems like a really convoluted way to do things," I noted.

"Brains are weird," Duchaine replied.

The Saturday after Thanksgiving, I'm hanging out with my dad's side of the family in a rented beach condo. We've just finished a big lunch of leftovers, and now we're all sleepy. My grandpa claims an empty bed, but Dingfelders are not big on personal space, so my dad and I join him.

"Dagnabbit," Grandpa mumbles. But he falls asleep before he can lodge any more complaints.

"How was your trip to Sarasota?" my dad half whispers. He's looking to gossip about his former in-laws, and I'm here for it.

"Uncle Aaron was there," I say. "I think he's living in his RV in Grandma's backyard, and the neighbors don't seem super keen on the situation."

"There's no way she's zoned for that," my dad says.

"It looks like Aunt Karen has an RV back there too," I add. "And I think they were trying to tie in to the city sewage line."

"No!" my dad exclaims, delighted.

"CuggghhhhhNORK," adds my grandpa, who has started to snore.

"The craziest thing happened," I continue. "I thought, for a second, that Aunt Karen was my mom!"

I fish my phone out of my pocket to show him a picture.

"She dyed her hair and now they are basically twins, don't you think?"

My dad squints at my phone and says nothing.

"Don't you think they look the same?" I prompt.

"I think Karen looks better with dark hair," my dad says, missing the point completely.

A few quiet seconds pass. I want to tell my dad about the study I'm participating in, but I'm not sure how it will go over.

"Have you ever heard of faceblindness?" I say. "It's this thing where you can't recognize people. I think I might have it."

"There's nothing wrong with you," my dad says with a laugh. "You're just spacey."

"It's a real disorder," I protest. "I'm actually going to go to Harvard to take a bunch of tests to see if I have it."

My dad says nothing.

"I think it might explain why I had so much trouble making friends when I was a kid," I add. "Though that wouldn't explain what changed when I got to college..."

I wait for my dad to chime in. And eventually he does.

"CuggghhhhhNORK," he says. He's fallen asleep even faster than Grandpa.

MISSED CONNECTIONS

AFTER LANDING IN BOSTON, I HEAD TO THE COFFEE SHOP where Pam is wrapping up a meeting with her writing group. Just being here makes me feel smart—there are floor-to-ceiling shelves crowded with leather-bound books, a crackling fireplace, and students everywhere. It reminds me of college—and Pam's abrupt, cackling laughter completes the effect.

As we journey to Pam's house, we catch up and gossip—and Pam has some top-shelf tea: a college friend of ours recently ran off with another college friend's wife!

"Wait, so who is [redacted]?" I ask. "Was she the tiny blonde who transferred to Brown?"

"No, that was [redacted]," Pam says. "[Redacted] lived on the third floor and brought her own horse to school."

We go back and forth until Pam gets out her phone and shows me a photo of a woman who doesn't look the least bit familiar.

According to Pam, this is someone I ate dinner with, every night, for two years.

"Oh yeah, *now* I remember," I lie.

College was a long time ago, and it's normal to forget people you used to know. But the fact that I hid my memory lapse means that I'm embarrassed by it. I should remember this gal.

When we get to Pam's apartment, her boyfriend lets us in.

"Hi..." I say.

I totally recognize him. I know what he does for work (computer stuff), and I remember the time he made an oil lobbyist cry at a friend's Christmas party. But I can't remember Pam's boyfriend's name.

This isn't faceblindness—it's lethonomia, forgetting names, and it happens to everyone on the order of once a week.[1] The problem is twofold: modern society requires us to remember far more names than we evolved to keep track of, and names are relatively meaningless. If someone tells you that he is a baker, you might think about bread and flour and waking up early—a rich array of baking-related targets that you can associate with a face. If someone tells you that his name is Baker, that's an immediate cognitive dead end.

Pam's boyfriend's name, however, turns out to be the tip of the iceberg. As we wait for the pizza we ordered to arrive, I find myself grasping for something to talk about. If only I could remember where Pam works, or what she does. Asking about someone's family is always a good strategy, but I can't remember if Pam has siblings, or anything about her parents—if they are still together, or even if they are still alive.

It's only a matter of time before I reveal my massive ignorance about every aspect of my friend's life. At one point, I ask for some water from Pam's boyfriend—a situation that obviously calls for a proper noun. "Hey, um, can you . . ." I say. It's so awkward! I should have just gotten it for myself.

I mentally run through my list of name-getting gambits. I could bring up how terrible my driver's license photo is and ask to see his. Another good trick is to bring up naming conventions in other countries. In Bali, for instance, people are named according to birth order. Firstborn sons are Wayan, Putu, or Gede, while firstborn daughters are Ni Luh. There are designated names for children two through four, and then at five you go back to Wayan, or Wayan Balik, which roughly translates to "Another Wayan." If only Americans would take up this tradition—it would give everyone much better name-guessing odds.

I'm just about to bring up Bali when I see my life raft. Peeking out from underneath a postcard from Bed, Bath & Beyond is an envelope addressed to Pam and David.

David! How am I going to remember that? Generic guys' names are the worst.

Before I go to bed that night, I trot out my hard-won knowledge. "Good night, Pam. Good night, *David*," I say.

Their guest bedroom doubles as a server room, so I'm bathed in green LED light as I tap out an email to myself on my phone. Subject: "Pam's partner's name is David." Then I tag it "People to remember to invite to things."

I invented that tag ages ago for its stated purpose, but it evolved into a repository of assorted information about my friends. It's full of blank emails with subjects like "Abigail has two sisters, one in London and one in Arlington," and "Serena's dog Bogo is getting knee surgery next week." I consult it before meeting up with people I haven't seen in a while.

Is this a workaround for a memory deficit I didn't even know I had?

Is it weird that I know a variety of ways to get people to say their own names?

I'm wide-awake now, thinking about my many life hacks—tricks and tips that I generously share but that never seem to catch on.

Once, in middle school, my best friend, Tracy, asked me if I had a spare maxi pad, and I was thrilled with the opportunity to share my workaround. Feminine products present so many problems: it's embarrassing to ask your parents to buy them for you, it's impossible to remember to have them on hand when you need them, and if they fall out of your backpack in school, you'll never hear the end of it.

Why bother with all that when... (drum roll)... wads of toilet paper work just as well!

"They are better for the environment, and free," I said.

I offered to show Tracy my wad-making technique. When she declined, I figured she simply lacked an appetite for innovation.

As I listen to the quiet clicking of David and Pam's hard drives, I realize something: the reason no one ever adopts my life hacks is because *they don't need them*.

Huh.

I'm at the Jamaica Plain VA Medical Center in Boston, in a small room with carpeting the color of old pencil erasers. I was dismayed, this morning, to discover that my testing would not be taking place within Harvard's ivy-covered embrace. As it turns out, DeGutis has a joint appointment at the Boston VA, so here we are.

Research assistant Alice Lee plops a chunky gray laptop in front of me, and I dive into a heap of tests. There's the Cambridge Face Memory Test again, with its collection of anhedonic frat boys. Then there's the Benton Facial Recognition Test, which asks me to match people who have been photographed at different angles and in extreme lighting conditions. Another test gives me a glimpse of disembodied lips, eyes, and noses and then asks me to identify which face they've landed on.

All these tests are hard, but I eventually get to one that is so difficult, I don't even understand *the directions.*

Does this make sense to you?

Put these six faces in order from most to least like the target face:

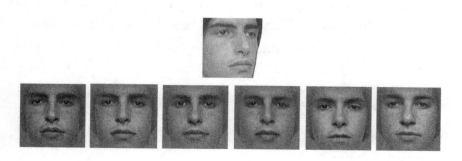

I feel like I've been asked to sort a pile of fruit from the most to least extraverted. It's absurd.*

"I don't understand what I'm supposed to be doing," I whine.

"They are challenging tests," Alice says evenly.

Toward the end of the session, I take a bathroom break. Alice shows me the way there, but I get lost on my way back. Without even thinking about it, I start following the "always turn left" maze rule.

On my meandering tour of the entire floor, I realize that being lost is so normal for me that I hardly even notice it anymore.

---

* Surprise! They were already in order. In the actual test, they are scrambled.

Come to think of it, I am weirdly tolerant of all kinds of ambiguity. I regularly find myself in situations in which I don't know where I am, what exactly I'm supposed to be doing, or who I'm talking to — situations that other people find quite alarming. I remember Steve getting all stressed out when I admitted, the morning of our wedding, that I did not know, precisely, where the ceremony was going to be. "Will you just take a breath and *chill out*?" I said. What kind of control freak was I marrying?

Agnosia, not knowing — that is sort of my specialty. I'm the most agnostic person I know.

A young woman peeks out of a door and calls my name. She has blond hair, while Alice had dark hair — at least that's what I remember. I could be wrong.

"Sadie?" she says.

"Hey!" I say warmly, as if we're old friends.

"I'm Anna," she says. "Alice had to go to class, so I'm taking over for her."

Well. That was tricky.

"I'm sorry that took so long," I say. "I got lost."

"Oh yeah, that's really common with..." she says, trailing off.

"With faceblind people?" I ask.

"With lots of people," Anna says. "It's a confusing layout."

Anna was just being kind. I later find out that topographical agnosia co-occurs with faceblindness about 29 percent of the time. It's probably related to the fact that the FFA is right next to the part of the brain that recognizes familiar scenes, the parahippocampal place area.

Neurodevelopmental disorders are like mice — for every one you notice, there are dozens more hidden underneath the floorboards. Bill Choisser, for instance, also had an auditory processing disorder that made it hard for him to decipher speech. Faceblind artist Chuck Close also had dyslexia. Some 36 percent of autistic people also have prosopagnosia[2] — though researchers often exclude autistic folks from faceblindness studies because that version of prosopagnosia may have a distinct developmental trajectory. (Whether that's an accurate assumption is up for debate.)

Among people with acquired prosopagnosia (due to a brain injury), there are at least two subtypes: apperceptive and associative. Apperceptive prosopagnosia happens at the first step in face processing, when your neurons are encoding a face, turning what you're looking at into a meaningful pattern of firing. This ability is measured by the Cambridge Face Perception Test — the face-ordering test with directions that didn't even make conceptual sense to me.

Even if you can encode faces, you can still get tripped up at the next step, which involves retrieving the stored image of a face and comparing it to whomever you are looking at. This ability is measured by the Cambridge Face Memory Test. If you flunk the CFMT but do well on the CFPT, you have associative prosopagnosia.

Apperceptive prosopagnosia tends to result from injury that is toward the back of the brain, while associative prosopagnosia generally results from injury to parts of your brain that are closer to your eyes. This is because after visual information leaves your eyes, it goes all the way to the back of your head, to what's known as your occipital lobe. ("Occipital," in the word's original Latin form, means "back of the head.") From there, visual information inches forward, toward your eyes. At each step, your mental representations become more and more sophisticated — from dots to lines to things to concepts.

Another potentially useful categorization is based on the nature of a developmental prosopagnosiac's (DP's) perceptual deficits. Neurotypical folks are much worse at identifying faces when they are upside down, as opposed to right side up. This, the so-called face-inversion effect, is generally taken as evidence that most humans see faces holistically, rather than as a collection of parts. A recent study found a subset of DPs who are immune to the inversion effect.[3] They are equally good (or bad) at identifying faces, no matter what the orientation—which suggests they are looking at features one by one, and not paying attention to how they hang together.

Generally speaking, if you want to know what faces look like to faceblind people, just find a picture of a famous person and turn it upside down. The features will remain perfectly sharp, but they probably won't hang together for you like they did before. You may even find that a celebrity whom you can easily identify right side up becomes unrecognizable when flipped upside down.

The implications of this are self-evident. If you're famous and want to travel incognito, try walking on your hands.

There's a nightmare-inducing twist on the face-inversion effect that was discovered by University of York psychology professor Pete Thompson in 1980. Thompson needed a printout of a famous face for a class demonstration. He dropped by his local conservative party office and nabbed leftover posters from Prime Minister Margaret Thatcher's recent, successful campaign.

When Thompson got home, he cut out her eyes and mouth and turned them upside down. The effect was delightfully ghoulish. Then he went to another room to grab some tape, and upon reentering, he was surprised to see that Thatcher's doctored face looked...kinda normal. At least, that was the case when he viewed it upside down. When he looked at it right side up, the Thatcher poster became freaky again.

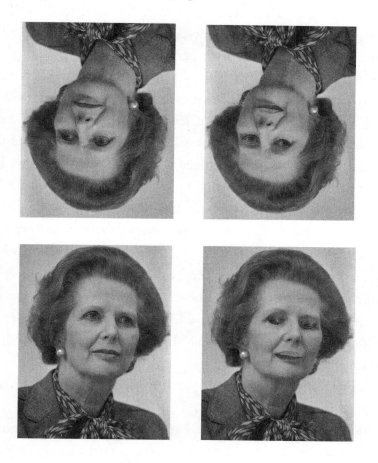

This, the "Thatcher Illusion," shows that humans are so good at making sense of right-side-up faces, we immediately spot any weirdness.[4] But when we look at upside-down faces, we're slower to see that something is amiss—even something as obvious as upside-down eyes.

Two hilarious rumors have circulated since Thompson published his discovery: that he defaced Thatcher because he was upset that she was making deep cuts to university budgets, and that this effect works only with her face. Unfortunately, both are apocryphal.

"I didn't like Thatcher, but that has nothing to do with the effect. It was purely serendipity," Thompson told me. "And, of course, it does work with other people."

In my rush to get to the VA hospital this morning, I failed to drink my customary gallon of iced tea. As a result, I've had a migraine brewing all day, and it decides to assert itself fully on my journey back to Pam's. I manage to find the right bus without fully opening my eyes. A woman sits next to me, and I can't help but notice that her face is slack, like an ill-fitting mask. On the other side of the aisle is an elderly man with caterpillar eyebrows and obscenely fleshy lips. When I glance at a perfectly lovely child and find myself startled by her freakishly big eyes, I realize I'm the problem, not them. You know how when you say a word over and over again it starts to sound like nonsense? I think that's like what's happening to me. I spent all day looking at faces, and now all faces look weird.

It's unsettling, but hey—at least my fellow passengers haven't swapped faces with their dogs.

When I turn my phone on, I see it's full of messages from the Anns. They both live farther out in the suburbs than I realized. Getting to them means taking an infrequent commuter train, and it's not clear if either of them has time to pick me up. It's starting to look like seeing them is not in the cards.

Honestly, though, I might be using logistics as an excuse. I can't remember a thing about their lives. I don't know their jobs or their kids' names, or even how many kids they have. What would we talk about? It's not like we can reminisce about college—I've forgotten that too!

What is up with my memory? Is this a faceblindness thing, or something else? I make a mental note to ask DeGutis, which, of course, I won't remember, so I write it on my hand.

When I get to Pam's, I take two Tylenols, chug a 20-ounce bottle of Diet Coke, and take a nap. An hour later, my migraine is gone and my life outlook is much improved. When I check my phone for breaking Meghan Markle news, I'm delighted to see that human faces no longer look like gargoyles to me. Phew.

But now I'm curious: *Are* human faces weird?

I go down an internet rabbit hole and emerge with an answer: yes.

Other animals must think we look so freaky. We don't need the strong jaw muscles that keep fox faces taut, so our mouth area is slack and rubbery—a necessary adaptation for talking. While other animals have uniformly dark eyes, ours show a lot of white so that we can infer where other humans are looking.[5] And then there are our furry, mobile eyebrows—which we may have evolved to amplify the movement of tiny face muscles to communicate emotions.

But you want to know the weirdest thing about human faces? They are each unique.

Some of us have little button noses, while others have big Roman honkers. Lips range from nearly invisible to plump Cupid's bows. Eyes run from deep-set to bulging, and all these features can be scrunched together tightly or spread out wide.

That's a lot more variety than you see in other animals' faces, and I'm not being speciesist. Penguins really do all look alike, even to each other.[6]

Interestingly, ancient hominids also had homogenous faces. It wasn't until about six hundred thousand years ago that faces started getting distinct.[7] Why? We don't know, but one hypothesis links it to the development of a new type of social structure, known as fission–fusion. In fission–fusion societies, group membership is flexible and ever-changing, requiring animals to keep track of scores of relationships over long periods of time. For instance, if a close ally returns after a lengthy absence, it's in both of your best interests to pick up your friendship where it left off. The evolution of highly individualized faces, as well as a specialized brain module devoted to remembering them, may have helped early hominids navigate an increasingly complex social world.

Or maybe we just developed a taste for the unusual. Researchers have found that if you show men a slideshow of women that mostly consists of blondes, the rare brunette gets a ratings boost, and vice versa.[8] This trick also works with women rating men who are clean-shaven or bearded.[9]

Hot bearded guys aside, however, human faces are substantially more hairless than other animals' faces. This may be, in part, because we use our faces to do a job that was previously handled by our butts: flushing to show sexual interest. Indeed, many primate butts both flush and swell to indicate they are ready to mate. Perhaps, as our ancestors took to walking upright, that signal had to migrate upward to be more conveniently at eye level, says Mariska Kret, a psychology professor at Leiden University.

Many primates can identify other individuals just by looking at their butts—and Kret believes that butts may have preceded faces as the best way to identify your friends. If so, that means *human faces evolved to resemble monkey butts*. (Or, more accurately, protohuman butts.)

If you don't see the resemblance, that could be because you haven't spent enough time looking at the nether regions of your fellow apes. "They both are always on display, they both flush, and they have distinct features that are bilaterally symmetrical and high contrast," Kret says.

Kret and her colleague Masaki Tomonaga have gathered evidence to support their hypothesis: chimps see butts in much the same way humans see faces—holistically.[10]

For their experiment, Kret and Tomonaga took pictures of human and chimp "anal-genital areas," feet, and faces. Kret handled the human photography, which involved finding highly depilated female volunteers. ("I can't believe we got IRB approval for this," she told me.) Tomonaga handled the chimp photography.

Next, the scientists presented the chimps and humans with one of these body parts (with each species seeing its own parts) and asked them to find the identical one on a touchscreen.

Both species were excellent at matching faces and not very good at matching feet. But what separated the chimps from the humans was the butt-matching task—the chimps seriously outclassed their less hairy competitors. What's more, the chimps experienced a butt-inversion effect, whereby they had more trouble matching upside-down butts than right-side-up ones. Humans were equally slow at butt matching, regardless of the orientation.

"Chimps are using holistic processing rather than going feature by feature, which is also how humans process faces," Kret says.

At some point, our primate ancestors lost their special butt-identification hardware, but we remained face-processing experts—and the foundation for this ability is built into our brains. Researchers shining lights on the bellies of pregnant women have found that eight-month-old fetuses orient to and track lights that make a face-like pattern (two dots over a line), and ignore that same pattern upside down.[11] This suggests that we are born with a face template ready to go.

Refining our face-identification skills, however, takes a lifetime of practice. Six-month-old infants, for instance, are just as good at identifying human faces and chimp faces. By nine months of age, they've lost the ability to distinguish between chimp faces, but they've improved in their human face-identification skills.[12] This perceptual narrowing also occurs along racial lines. Newborn infants are equally good at learning the faces of people from different races, but by three months, they begin to have trouble distinguishing between the types of faces they see less often.[13] This is the beginning of the "other-race effect," whereby people think (but hopefully don't say) that people from other races all look alike. Scientists have found that the effect can be reversed by simply showing infants photos of different kinds of faces.[14] However, to fully defeat the other-race effect, you may have to keep this up until your baby is well into their thirties—which is when human face-identification skills peak.[15]

Our inborn face template gives us a major head start on the subtle art of face identification, but it also results in some strange mistakes. If you've been alarmed by the screaming face inside a bell pepper, or felt silently judged by your grim-faced electrical outlets, you've experienced one such phenomenon, known as face pareidolia. From an evolutionary perspective, it's much worse to miss a face than to see a face where none is present—so our visual systems err on the side of caution. Interestingly, these phantom faces activate our FFAs much like real faces,[16] and we automatically attempt to glean how they are feeling.[17]

If humans experience face pareidolia, does that mean chimps experience *butt* pareidolia? Do they see primate nether regions in passing clouds or tangles of vines? When I asked Kret, she just laughed. I take that to mean: yes, they definitely do.

I'm back on the eleventh floor of the hospital, but I'm not lost this time. I'm waiting for either Anna or Alice to pick me up, and I don't quite remember what either of them looks like. There aren't any chairs, so I sit on the floor and beam a warm smile at every young, long-haired woman who passes by. One of them eyeballs me warily. I wonder how many other people I've creeped out.

After ten minutes, I email Anna, who tells me that I'm on the wrong floor. When I find her on the twelfth floor, we walk down the hall to my appointment with Joseph DeGutis, ~~my torturer~~ the lead scientist on this study.

DeGutis, a young, athletic man, stands up from his chair and thanks me for coming. I know I'm going to have to find him later, when we meet up for my fMRI brain scan, so I try to memorize his face—handsome, if a little wolfish.

"So, am I faceblind?" I ask abruptly.

"We think you have mild to moderate prosopagnosia," he says. DeGutis doesn't want to elaborate because knowing too much about the experiment could taint my data. "We'll tell you everything you could ever want to know when you're done."

As we chat, I begin to understand why there's so much research on prosopagnosia. Most scientists are not in it to cure or even treat the disorder. Rather, they are hoping to get a handle on the much bigger question of object recognition.

Consider the surprisingly undefinable category of chairs. There's so much variety! Some chairs are light and portable; others are huge and heavy. Features like arms and headrests are common but not compulsory.

And yet, we know a chair when we see one. Do our neurons encode some sort of platonic ideal of a chair? Is that chair the average of every chair we've ever seen? How do we identify chairs from various viewpoints, or in weird lighting conditions?

Many prosopagnosia researchers would just as soon study chair identification, but until recently we had no idea where in the brain that takes place.* We do, however, know where in the brain face recognition happens: in the right hemisphere's FFA. With this information in hand, neuroscientists can do nifty things like eavesdrop on individual FFA neurons and record how they react to various kinds of faces.

"From a scientific level, it's one of the most specific things going on in the brain that we know about," DeGutis says.

While most face-recognition research is done in the name of basic science, a few potential treatments have popped up along the way. DeGutis, for example, was studying how people distinguish between similar faces when he realized that he could turn his experimental stimulus—computer-generated faces with subtle shifts in feature spacing—into a training program. He tested it on a friend of a friend and was stunned when it worked.

"I was like, 'Oh my God!'" DeGutis recalls. "'We might have actually helped someone!'"

As the interview wraps up, I blurt out a question that I hope sounds entirely theoretical.

---

* Doris Tsao, whom we will meet later, is working on this problem.

"Does remembering faces have any relationship to remembering *people*?" I ask. "Like, do other faceblind people have trouble remembering the names of their friends' boyfriends?"

"That's not something a lot of people complain about," DeGutis says. "But there is some evidence that faces act as a node for other biographical information—like where someone grew up and what they studied in college. Things like that might be hard to retrieve if you can't recall someone's face."

This is fantastic news. Having a neurological disorder is much better than being an asshole.

While I am increasingly aware of the many things that I am not good at, there is one area where I really shine, and I'm excited to show it off before leaving Boston.

"It's very important that you stay completely still," the fMRI technician says. "No fidgeting."

"No problem," I say. Not to brag, but I'm excellent at doing nothing.

I lie down on a gurney, which then slides into a dark tunnel.

"Are you all right in there?" the technician asks. I briefly squeeze a ball to signal that I am fine.

My job is to watch a movie, and it's not very interesting. There's no story, just a series of shaky clips: a child's elbow, a journey down a country road, a ball bouncing in a parking lot, a doughy leg, half a face. The fMRI machine provides a suitably avant-garde soundtrack of clicks and whirrs.

A tinny voice startles me back to consciousness: "Squeeze the bulb if you're still awake." I spend the rest of the ninety-minute session digging my fingernails into my palm to stay alert.

Finally, the movie ends and I'm rolled out of the machine.

"You were completely still the whole time," the technician exclaims. "I wish everyone were as good as you!"

"It was nothing," I say modestly.

After the fMRI, I take the T to Logan Airport, get on a plane, and fly back to DC.

"Why'd you come home early?" Steve asks. "Weren't you going to see a bunch of your old friends?"

"Oh, I don't know," I sigh. "I guess I just wasn't feeling it."

HACKING THE SYSTEM

The Monday after I get back to DC, I'm in the CVS and a familiar-looking woman smiles at me from the greeting-card aisle. "Hi," I say. "I haven't seen you in a minute." This is one of my favorite salutations, because it prompts people to mention the last time we hung out.

The card-buying lady looks right through me, so I pretend that I'm talking to someone behind her. "Let's catch up later!" I chirp.

The evidence that I have prosopagnosia is piling up, but I can't see it. Sure, DeGutis told me that I *might* have "mild to moderate" prosopagnosia, but isn't that just the low end of normal? In any case, I'll have to wait until the experiment is over to find out.

I round myself up to "basically normal," and this belief is further confirmed when I join a faceblindness Facebook group and find that it's full of stories of real suffering. There's a mother who lives in fear of picking up the wrong kid from school. Another mom failed to recognize her child's teachers, and now she's fighting rumors that she's a bad parent. There's a man who accidentally ignored his crush at a pub, and now she's giving him the cold shoulder. A woman whose cell phone was stolen couldn't pick the thief out of a lineup and became the town's laughingstock.

Most harrowing of all are stories from faceblind women who have *stalkers* they can't recognize. They describe being constantly on edge, unable to protect themselves from an effectively invisible menace. *What a nightmare*, I think. *Good thing I'm just a little below average.*

When I receive an email from Alice, inviting me to take part in DeGutis's face-recognition training program, I'm somewhat taken aback. Does this mean I'm legit faceblind? Maybe I'm just a control subject.

I'm probably just a control.

"Of course!" I write back.

Alice sends directions for logging on to the testing website from the comfort of my own bed. The instructions explain that I'll be shown a grid of ten faces—actually, variations on the same face, with some of the features shifted by tiny increments. A zigzag line separates these nearly identical mugs into two groups: in category 1, the faces have eyes and mouths that are spaced farther apart; in category 2, the faces have more compact features. My job is to remember which faces belong to category 1 and which belong to category 2. It's a weird activity, but I know from reading DeGutis's earlier work that it's supposed to teach me to focus on information-rich areas of the face and make fine distinctions on the fly.

After studying the grid for a good fifteen minutes, I hit START, and the faces begin flashing by. I am supposed to decide which

HACKING THE SYSTEM

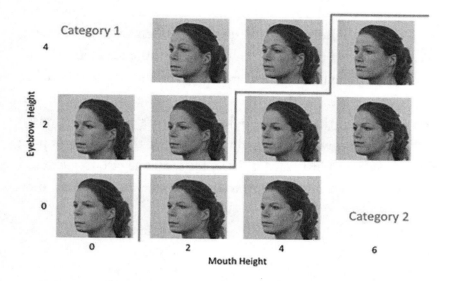

group each face belongs to in less than a second, and I am choosing completely at random.

For the second round, I try to measure the distances between features using my fingernails, but this takes too long. The screen flashes red and the faces disappear, which means I'm out of time. I'm back to guessing.

On the third round, things get even harder: The sizes of faces change. Some are as big as index cards; others are closer to postage stamps. To sort these faces, I have to remember the *relative* distance between the features.

"Somehow I'm doing worse than chance," I complain to Steve.

"That's *good*," Steve says brightly. "Just do the opposite of whatever you're doing now."

(I roll my eyes. Math people, so literal.)

It's past my customary 9 p.m. bedtime, so I chalk up my failure to fatigue and decide to try again in the morning.

On my second day of training, I name the two groups the Steves and the Bobs, and I attempt to attach feelings to them, because I've read that strong emotions can enhance memory. The Steves, with their compact features, seem smarter to me, more handsome. Steves do the dishes without you even having to ask. Then there are the boorish Bobs, with their perpetually raised eyebrows and low, simian jawlines. Bobs let their dogs poop on the sidewalk and just leave it there, for someone else to pick up. I stare at each face and attempt to drum up feelings of admiration or repulsion.

This strategy makes the task less boring, but my scores remain abysmal.

I email Alice, begging for help.

"Do I need to make and study flash cards before doing the computer tests?" I ask.

"Nope," she replies. "Please do not create any flash cards or do any other face studying...because it could become a confounding variable and bias your results." She also tells me to go faster — each session is supposed to take thirty to forty-five minutes, and I'm taking an hour.

This advice does not strike me as particularly helpful.

Five nights a week, I continue to run full tilt into DeGutis's brick wall. One evening, I stare wide-eyed at the grid of faces, not blinking, trying to burn them into my retinas. Another night, I give the faces backstories that explain the particular spacing of their features. Imagining gruesome industrial accidents takes the edge off my frustration, but it does nothing for my scores.

When I'm about three weeks in, I start to despair. I can't just keep failing the same test, night after night. I'd rather give my cats eye drops or do chess puzzles with Steve.

HACKING THE SYSTEM

Taking in the face holistically isn't working, and neither is scrutinizing its component parts. So, one night I split the difference and divide each face into halves, top and bottom. Then I decide which of three positions (high, middle, or low) describes the eyebrows, and I lightly tap "1" on my keyboard for high, "2" for low, or both buttons for middle. For the bottom half of the face, I tap 1 for a high mouth, 1 and 2 for a middle mouth, and 2 for a low mouth. The resulting tapping pattern tells me which kind of face it is: 1 and 1 = type 1, 1 and 2 = type 2, and so forth.

I'm amazed when this convoluted strategy starts to work!

At first I have to talk to myself while tapping, which slows me down. With practice, however, I'm able to go by the taps alone, and I start flying through the test. Weeks go by, the faces get harder, and my scores stay high. A few times, I come tantalizingly close to getting a perfect score.

Forget remediation—I want to become a face-recognition expert! The possibilities are endless: Me flagging down the correct waitress in a restaurant. Me talking to a coworker without having to check the cubicle sign first. Me recognizing someone out of context, and maybe even recalling their name.

As it turns out, there are people who can do this and more, and they are known as super recognizers.

When Andy Pope's face pops up on my computer screen, my finger instinctively hovers over the 2 button. His features are quite compact—eyes, nose, and mouth all centered on a perfectly oval head. His hair and beard are sandy blond, and he looks about ten years younger than he is.

"I get a lot of exercise in my line of work," Pope says modestly.

In his job as a community police officer in West Midlands, England, Pope spends much of his day walking around town and talking with citizens and shopkeepers. He meets dozens of folks every week, and he pretty much remembers them all.

"Sometimes I forget a name, but I usually do remember people's faces," he says.

I ask Pope what it's like to be him: Does he love going to parties? Is he free from social anxiety? Does he ever think about running for office?

"No. I'm quite shy, actually," he says.

Being a super recognizer isn't all upside, says Josh Davis, a psychology professor at the University of Greenwich. Pretty often super recognizers will recognize a person whom they really shouldn't know—for instance, the younger sister of their fifth-grade best friend. In these cases, they feel compelled to pretend they haven't recognized that person. "They don't want to seem creepy, or like a stalker."

Pope's main complaint is that the face-matching part of his brain never goes off duty. A few years ago, on his fortieth birthday, Pope was having a beer in Birmingham city center when he saw a middle-aged man who was wanted for a series of incidents "of a sexual nature" aboard buses. Pope had never seen the guy in real life before, but he had seen videos of him from CCTV cameras.

"Luckily some members of my team were on an operation that night, so I called one of them and he took it from there," Pope says.

Police forces around the world are beginning to take advantage of super recognizers, says Mike Neville, a former detective chief inspector at New Scotland Yard. "They are the biggest advance we've had in forensics since DNA testing," he says.

HACKING THE SYSTEM

In 2010, London security cameras were capturing tons of crimes, but very few suspects were ever identified or arrested.

"The original idea was that the cameras would themselves act as a deterrent, but it didn't work out that way," Neville says.

To put names to faces, Neville began circulating still images of suspects to police stations around the city. Identifications began to roll in, and an interesting pattern emerged. Out of a police force of more than thirty thousand, just twenty-five officers were making more than half of the IDs.

"At the time, I just chalked it up to motivation," Neville says.

Neville met Davis at a conference that year, and the two of them decided to put Neville's top six officers to the test—actually, a battery of tests, including some of the ones I took. The officers all did extremely well, and one guy got *every* question right on the Cambridge Face Memory Test.

"They were hitting the ceiling, so I knew I'd have to go back and make harder tests," Davis says.

With this data in hand, Neville created the New Scotland Yard Super Recogniser Unit. The team has since chalked up some major victories. They solved the 2014 murder of a London teenager, and in 2018, they identified two men who poisoned former Russian military intelligence officer Sergei Skripal.

While the super recognizers are sometimes deployed to scan crowds for wanted criminals, the unit spends most of its time sifting through security camera footage, looking for repeat offenders—people like Alexander Caballero, who stole more than £100,000 in goods from high-end stores in London from 2013 to 2015. A super recognizer spotted him in forty different security camera videos, pocketing luxury items like diamond bracelets, beauty cream, and cashmere scarves, and she flagged his record.

Caballero remained at large until January 1, 2015, when he was arrested after refusing to pay his cab fare and then hitting the driver with a shoe. At the police station, Caballero used a fake name and claimed never to have been arrested before. When the police fingerprinted Caballero, they saw that he was wanted for a truly epic amount of shoplifting. Within hours, the thief was presented with a compilation of his greatest hits, and he pleaded guilty to them all.

"If it weren't for the super recognizers, he would have only been charged with the assault," Neville says.

Neville now runs a consulting firm, Super Recognisers International, which helps law enforcement agencies identify super recognizers who are already in their ranks. He also runs training sessions to teach super recognizers how to best use their superpowers in particular situations, such as scanning crowds for known terrorists.

What Neville can't do, however, is teach neurotypical people to be super recognizers. This ability results from a rare mix of inborn capability and early life experience, Davis says. "Either you have it or you don't."

If these superheroes have a kryptonite, it's the face-inversion test. When asked to match upside-down faces, super recognizers show a particularly steep decline in their speed and accuracy.[1] It would be hard for a criminal to put this to practical use, but it does provide a clue as to the source of super recognizers' superpowers. Perhaps they see faces even more holistically than neurotypical people do.

Another clue comes from eye-tracking studies. When trying to learn a new face, most people's eyes bounce around, fixating on multiple points. Super recognizers, on the other hand, focus more steadily. They fixate on a point in the center of the face, which suggests they are able to take in the entire face at a glance.[2]

"It's the one-shot learning. They need very little exposure and information to form a representation that is absolutely richer than those that we generate," says Meike Ramon, a University of Lausanne professor who studies super recognizers.

"Is there anything that they are all really bad at?" I ask hopefully.

"No, there's no apparent cost to them cognitively," Ramon says, sounding a little disappointed herself.

Interestingly, they do not do particularly well on other kinds of memory tasks. Pope, for instance, forgets birthdays and anniversaries, and he never goes grocery shopping without a list.

And sometimes—rarely, but sometimes—he can't place a face.

"One time, I'd been out with a few friends. I was walking back home on quite a dark sort of footpath round the side of a lake, and somebody was walking towards me... and they knew exactly who I was, and I was thinking, *I've got no idea who you are*, so I just kind of nodded and smiled."

(*One* time?! This happens to me once a day!)

"Gosh, that sounds stressful," I say, my voice a little flat.

"It was so embarrassing," Pope says, and I must admit, he sounds truly rattled. "Still to this day I don't know who it was."

I'm murmuring to myself and stabbing at my keyboard one night when Steve asks a good question: "How long do you have to do this face-training thing?"

"Forever, probably," I reply. Even if I improve, the effect will fade if I don't continue to do occasional booster sessions.

"I mean, how long do you have to do this *tonight*," Steve clarifies. My muttering is getting in the way of *his* self-improvement project: rewatching every episode of *Brooklyn Nine-Nine*.

One evening, toward the end of my training program, I run into my friend Dani in the Metro. She's wearing a knit cap that covers her adorable ringlets of hair, but her heart-shaped face and deep dimples give her away.

I'm so proud of myself for recognizing my friend—a friend who is *wearing a hat*!

"I think this face-training thing is actually working," I crow.

Dani opens her mouth, but no sound comes out.

"What?"

"Well, the other day, you sat right next to me in Peet's, and..."

I know the rest of this story: I treated her like a total stranger. I've told just about everyone—friends, acquaintances, baristas—about my prosopagnosia story, and they've all started telling me when I accidentally ignore them. This happens a lot more often than I realized.

"Why didn't you say hi?" I ask.

"You looked busy," she explains.

Anger and jealousy well up in my chest. It's not fair! Why does Dani get to recognize other people effortlessly? Why do I *still* suck at it, even after devoting hours and hours to the project?

I'm beginning to suspect that my improving abilities aren't transferring to real life. All I've learned is how to beat the test.

Later that week, something happens that makes me feel a little less pessimistic. An annoying neighbor of mine walks into a coffee shop, and I quickly avert my eyes. I snubbed someone *on purpose*, for perhaps the first time in my entire life!

This is the first thing I tell DeGutis when I return to Boston for follow-up tests.

HACKING THE SYSTEM

"I thought that I hadn't learned anything, but then I ran into a guy who always wants to talk to me about his ferrets, and I avoided him," I boast.

"That's great!" Joe exclaims.

I'm hit with an unexpected wave of regret. Do normal people regularly pretend not to see one another? And do I really want to join them?

YOUR BRAIN'S ROSETTA STONE

Imagine, if you will, that you're having brain surgery tomorrow. You're lying in a hospital bed with a swirl of fabric around your head and a bundle of wires coming out of the top. A week ago, neurosurgeons poked electrodes into your brain in the hope of finding the source of your seizures. But today, with your permission, a group of affable scientists are taking advantage of your strange situation to... Well, you're not exactly sure what they are up to. All you know is that they interviewed you about your interests, and now they are making you watch a weird slideshow. You're supposed to click a button every time you see a human face, but you know that's just a ploy to keep you focused. The real data is coming directly from individual neurons in your brain. The wires coming out of your head wrap are transmitting their signals to a computer across the room, which displays each neuron's response as a line on a multicolored graph.

Today's slideshow is typical. A variety of different pictures flash by quickly: Kobe Bryant, a spider, the Sydney Opera House, Jennifer Aniston, a dolphin, the Twin Towers, Jennifer Aniston with Brad Pitt, Sarah Michelle Gellar, the words "Jennifer Aniston" written out, the Leaning Tower of Pisa...

What's with all the Jennifer Anistons today? You know the scientists won't tell you the answer. It would mess up their data.

In July 2005, a team of scientists discovered a "Jennifer Aniston neuron" in the brain of an epilepsy patient. It fired in response to all sorts of different pictures of the actress and to her written-out name—and it ignored other, similar-looking actresses.[1] In another patient awaiting surgery, the scientists found a neuron specific to Halle Berry—and it fired in response to the actress even when her face was covered with her Cat Woman mask.

Was this evidence that we assign a single neuron to every person in our life? When I pitched this story to the science experts at the American Psychological Association, they scoffed. Something as complicated as your concept of another human being is going to be distributed across networks of neurons spanning huge swaths of your brain. "The 'grandmother neuron' hypothesis is a joke, not a serious theory," one of them said.

And that was that. I hadn't thought about the Jennifer Aniston neurons for thirteen-odd years, until I started investigating faceblindness. Are malfunctioning Jennifer Aniston neurons to blame?

I call up the original study's lead scientist, Rodrigo Quian Quiroga, to find out. And he immediately tells me something surprising: the APA science experts were right.

"Everyone was saying we'd found a neuron that represents a single person," says Quian Quiroga, now director of the Centre for Systems Neuroscience at the University of Leicester. "But what we found were neurons that represented particular *concepts*."

The Jennifer Aniston neuron—which was located in a deep brain structure known as the hippocampus—didn't just fire in response to Aniston. It also fired in response to Aniston's *Friends* co-star Lisa Kudrow. And it refused to respond when Aniston was pictured with her real-life boyfriend, Brad Pitt. "We think it encoded the concept of her character on *Friends*," and not the actress herself, Quian Quiroga tells me.

In a follow-up study, Quian Quiroga and his colleagues paired pictures of Aniston with pictures of the Eiffel Tower, and they watched as the so-called "Aniston" neuron began firing in response to the Paris landmark.[2] Quian Quiroga believes that this shows the "Aniston" neuron is a way station involved in linking associated concepts.

If I want to know how the brain recognizes faces, Quian Quiroga recommends I chat with his colleague Doris Tsao.

Before following his advice, I check out Tsao's research and find a groundbreaking experiment she and her colleague Le Chang did in 2017.[3] First, the scientists created computer-generated faces by systematically manipulating fifty variables, such as the distance between the eyes. Then they showed macaques the faces while recording the activity of about two hundred individual neurons in the face-recognition areas in each of their brains.

Tsao and Chang checked their work by showing the macaques a new face—one the monkeys and the researchers had never seen before. The monkeys' neurons fired away, and a computer armed with Tsao's face-generation formula spat out a picture that was nearly identical to the one the macaques had been looking at.

DO I KNOW YOU?

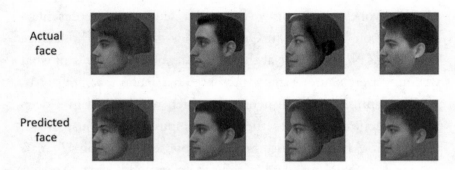

Actual face

Predicted face

(Doris Tsao, California Institute of Technology)

"You know how you say 'A picture is worth a thousand words'? Well, I like to say a face is worth about two hundred neurons," Tsao says.

It's looking like the brain encodes faces by mapping them onto a fifty-dimensional "face space," with an average face at the center, your personal platonic ideal. Small groups of neurons measure how far each new face you see deviates from this average along particular dimensions, such as eye width or skin texture.

Tsao's lab is now working out how the brain recognizes other kinds of objects, and they are creating a map that predicts the location of neurons that encode things from chairs to cats to body parts. The face-space system is serving as their Rosetta stone. "When we figure out the code for faces, then we can figure out the code for all objects," she says.

So far, it's looking like the IT cortex starts by asking two basic questions: (1) Is this thing animate or inanimate? (2) Is it blobby or spiky? (Faces, obviously, land in the blobby-animate quadrant.)[4]

This may seem like an odd way to sort things, but it must work well, because some of the most advanced, object-identifying computer programs have converged on this exact solution.[5] All you have to do is feed these programs lots of photos. You don't even have to tell them what they're looking at. These neural nets figure it out on their own, and—like the IT cortex—they start by sorting objects into the categories of alive vs. not alive, and spiky vs. blobby.

Who do we have to thank for this major finding? I mean, besides Tsao and other brilliant, hardworking scientists? Prosopagnosiacs, of course! Tsao's work built on a foundation of knowledge dating back to Joachim Bodamer, which pointed to a specific place in the brain for face identification. Over many decades, scientists were able to pinpoint this precise location by collaborating with faceblind people who donated their time (and sometimes their actual brains) to science. So, people of the future: next time your robot dog successfully fetches your slippers, be sure to thank your faceblind friends. (The assassin robots, however, are not our fault.)

OBLIVIOUS INEPTITUDE

Right before I find out whether I am officially faceblind, I have an appointment with the *Washington Post*'s daily podcast team. Host Martine Powers does not mince words.

"What do you expect the doctors will say?" she asks.

"I think they're gonna say I'm moderately faceblind, but probably not enough to be official," I reply.

As Powers and I talk about the loneliness and isolation experienced by faceblind kids, I hear my voice waver. Uh-oh.

The podcasters send me to a different sound booth to call DeGutis, and I swipe a box of tissues along the way. Something tells me I'm going to need them.

"Hey, how—" DeGutis starts to say.

"So am I legit faceblind?" I ask, cutting him off.

"Yes. Your ability to learn new faces is among the worst of our prosopagnosiacs," he replies. "You were one of the lowest scores that we've registered so far."

Wow. I'm not just below average, I'm the worst of the worst!

"What did my brain scan say?" I ask. That, to me, seems like the big question. I could have (unconsciously) flubbed the tests in order to get a cool diagnosis. But even a trickster like me can't fool an MRI machine.

"So, this is really interesting," DeGutis says. "We found that your fusiform face area is actually thicker than average. Thick *sounds* good, but thinner is better for your face area."

Children start out with thick FFAs, but as the brain determines which neurons are useful and which ones are just getting in the way, it thins out the useless ones, creates more connections among the useful ones, and adds insulation to improve the connection speed. This so-called neural pruning and myelinization seem to have stopped short in my brain, at least in this one area.

"You have the fusiform face area of a twelve-year-old," he says.

"And the facial recognition ability of, what, a three-year-old?" I ask.

"I would say maybe like a mediocre or below-average macaque," he replies.

I laugh, but I feel numb, which is never a good sign.

"You were really on the low end of prosopagnosiacs," DeGutis continues. "You were really struggling, and the amount that you went up on these tests after training was kind of remarkable. If you'd been this good when you came in, you wouldn't have gotten into the study."

As I digest this, I feel a little burst of pride. I'm the best of the worst!

OBLIVIOUS INEPTITUDE

DeGutis and his colleagues were puzzled as to how I had managed this feat. The follow-up tests showed that my face perception is still terrible. I can't match a face with the exact same face at a slightly different angle or in different lighting, even with both faces right in front of me. But whatever sketchy information I managed to glean, I was now holding on to ferociously, through some unknown strategy or sheer force of will.

"Maybe you're doing more of this associative encoding and paying attention to kind of compensate a little bit more," he says. "But I don't think we fixed your perceptual deficit."

This tracks. If there's one thing I'm good at, it's finding clever ways around seemingly impassable obstacles.

For the first half of our conversation, I'm interested in a dispassionate way, as if we're talking about someone else. But then the numbness slowly starts receding, and I'm hit with a tsunami of conflicting emotions: sadness, relief, confusion, enlightenment, jealousy, vindication. I'm still talking, and I still sound normal, hopefully, but I'm simultaneously having a full-on meltdown. Tears and snot are streaming down my face — and my face is at *work*. The moment we hang up, I throw on my sunglasses, sneak out of the office, and go home to nap.

When I wake up, I call my dad with my news.

"So, it turns out that I am legit faceblind," I say.

My dad chuckles awkwardly. "Nah, your problem is that you just don't pay attention."

"No, really!" I say, frustrated. "I have a real neurological disability, and I've probably had it my entire life."

"There's nothing wrong with your brain," my dad replies. "And anyway, why worry about something you can't change?"

I hang up more confused than before. "Don't overthink things." "Fake it till you make it." "Stay positive and everything will sort itself out." This is my dad's approach to life, and it has always worked for me. Why am I questioning it now?

At dinner later, I tell Steve about my intense day. It's such a relief to learn that my problems remembering faces — and perhaps even remembering people — are the result of a quirk in my visual processing system, and not a character flaw. But I'm haunted by the possibility that I've just discovered a new word for the same problem. Maybe my thick FFA is *because* I'm self-centered or a bad friend. Maybe it didn't get the neural pruning it needed because I was too self-absorbed to bother looking at the people around me.

Where does my brain end and "me" begin?

Steve bites into his burrito and chews thoughtfully.

"Do you believe in a spirit that's separate from your body?" he asks.

"No," I concede.

"If you believe there is only matter, this is really just providing this super insight. It's not like your spirit has been fighting against this deficit. It's all the same thing, it's all you. You are an emergent property of your brain," he says.

This is a good point, but now I'm even more perplexed. Do I have a neurological disorder, or *am* I the neurological disorder? Are we one and the same?

For weeks afterward, I'm a mess. Absolutely anything can bring me to tears. I cry in a museum. I sob at my friend's kid's birthday party. I weep in front of a giant fish tank at the zoo. I sniffle dramatically when a school bus passes me on my bike. Eventually, the trigger becomes clear: middle schoolers.

This whole investigation has reopened a mystery from my past, one that I thought I'd solved long ago. Why didn't I have any friends when I was a kid? My old verdict—that I was weird and kids are mean—had suddenly been thrown into question.

In the spirit of exploration (why stop now, right?), I do something a little impulsive. I write a note, via Facebook, to my former classmates, telling them that I have just been diagnosed as faceblind, and asking if that had anything to do with my unpopularity.

"Do any of you remember anything relevant?" I ask. "I'm really just curious at this point. I'm a happy, successful adult now, and I have a thick skin. So if you recall that I was an off-putting weirdo, or that no one was friends with me because I smelled bad, I promise I won't be offended."

Replies roll in quickly. "You had friends! I was your friend!" writes my friend Saba.

Oh yeah! Now I remember. I loved Saba because she was smart and funny, but now I realize there was another aspect of her appeal. As the only Black girl in my grade, she was probably one of the few people I could reliably find on the playground. We lost touch in middle school—maybe because she was sorted into gifted classes and left me behind. Or maybe, as classes got bigger and schools more diverse, I simply couldn't pick her out of the crowd.

Other people tell me that I wasn't particularly unpopular, but I was rather standoffish. "You always seemed to just be content doing your own thing," one gal writes. "You always beat to your own drum," another says. A third classmate claims that she'd tried to be friends with me but that I never seemed to warm up to her. "We hung out a little and I thought you just didn't really like me," she says.

I'm floored, grateful, and heartbroken, all at the same time. I spent so many years drowning in loneliness, unable to see the life rafts that were being tossed in my direction.

A few days later, I receive an intriguing message from an old classmate. She says, "My cousin was obsessed with you. We couldn't figure out why you always ignored him."

I think I remember this mystery cousin—a blond guy who walked me to piano lessons a few times. I thought he was dreamy—but one day he didn't show up, and I never saw him again. I was disappointed, but not surprised. People disappear sometimes; I never thought to question it.

Is this why I have never, ever been asked out by a guy?

"Why do *I* always have to do the asking?" I once whined to my friend Sybil.

"It's because of your name," she replied. "You're like Sadie Hawkins."

It didn't help that I went to a women's college and sprouted a little neck beard that I failed to notice until my mid-twenties. But I suspect there were at least a few guys who would have asked me out anyway, whom I accidentally rebuffed or scared away with mixed signals, simply because I didn't recognize them.

"If I weren't faceblind, I would have been so popular—I would have been a cheerleader," I say to Steve one evening.

Steve laughs, and I punch him.

"Okay, not a cheerleader, but class president maybe."

The thing is, I *was* class president—well, not class president, but I was president of my dorm in college. My only opponent was Brown Anne, and I beat her in a landslide. Just a year earlier, I'd been utterly friendless and had a pile of transfer applications on my desk.

What happened? I think I've figured it out, and it involves

OBLIVIOUS INEPTITUDE

another grocery store epiphany. (What can I say? Some people have church; I have the snack aisle.)

I was nineteen, home from my first semester in college, and my dad seemed annoyed with me as we filled his truck with grocery bags.

"Well, that was rude," he said as we drove off.

"What?" I asked.

"Your buddy Susan," he said. "You just walked right past her."

"Oh, that was Susan?" I asked. A girl with short brown hair had waved to me in the store. "I said, 'Hi.'"

"You said, 'Hey-ay,'" my dad said, mimicking my limp, sing-song greeting. "You do that a lot."

"She's not my buddy," I argued. "I haven't seen her since middle school."

I later found out that Susan had continued with me to high school, and that she'd been a little hurt when I started treating her like a stranger. Something had happened—perhaps she'd gotten a haircut or glasses—that made me unable to connect high school Susan with my middle school friend. And when I stopped seeing her regularly, she simply faded from my mind.

Of course, I didn't know any of this at age nineteen. All I knew was that people I didn't recognize often greeted me, and I never stopped to chat for fear they'd figure out my ignorance and be offended. How, I asked my dad, do you have a conversation with someone if you don't know who they are?

"Everyone wants to talk about themselves," he said. "Just ask a lot of questions, and they'll think you're the most fascinating person in the world."

I brought this tip back to college, and it transformed my life. In one semester, I went from being lonely all the time to having friends spilling out of my dorm room. All it took was pretending to know the people who appeared to know me. When I was walking to class, if someone seemed to be looking my way, I smiled. If they smiled, I stopped to chat. Before long, the whole campus was brimming with close, personal friends. That I had no idea who they were seemed like a minor quibble.

There were a few people I could recognize. My friend Melissa had long blue hair; Thalia and Annette were tall, thin, and usually wet from swim practice. As for everyone else, I was often at a loss, but that didn't hamper my blossoming social life. When a vaguely familiar-looking woman flopped onto my bed and started talking about her love life, I just rolled with it. The trick was not to give away the fact that, as far as I knew, this was the first time I was hearing about this jerk of a boyfriend.

My popularity reached its peak when I was elected house president. I took pictures of all eighty women I lived with and put their faces and names on a bulletin board, claiming it was for everyone's benefit. I also kept a set of photos and wrote the names and hometowns of my classmates on the back. Despite nightly drilling with my flash cards, I never really matched faces with names. I did, however, link names to hometowns.

"And then Melissa used my hairbrush again..." the strange woman on my bed complained. "Melissa from New Hampshire?" I asked. There was no reason for me to bring that up, but I was so pleased with myself for remembering something, I couldn't help but blurt it out.

In the weeks following my diagnosis, I am suddenly, nakedly aware of just how bad I am at remembering faces. The extent of my ignorance is breathtaking. At a boring staff meeting, I tally the number of people sitting around the conference table who I can name. (Fewer than half!) When I'm walking around my neighborhood, I notice that I'm constantly stopping to chat with complete strangers who seem to know me. And despite this precaution, I regularly snub friends and acquaintances.

"You just walked right past someone who was trying to get your attention," my editor, David Rowell, says one day, just in time for me to turn around and see my boss disappear into a crowd.

"You should have said something!" I complain, though I know it's not fair. So many crucial social interactions happen in the blink of an eye.

Suddenly self-conscious, I catch myself avoiding eye contact and feeling anxious about parties. And I love parties! This is exactly what my dad warned me about. Can self-knowledge be dangerous?

To find out, I call William von Hippel, a psychology professor at the University of Queensland. He studies self-deception and has found that it can be quite helpful — especially in social situations. In one of his studies, participants were asked to review video footage of a guy named Mark and then write a speech about how he is either a great guy or a bad dude. Some of the participants watched videos that gradually went from positive (Mark helping someone in distress) to neutral (Mark making lunch) to negative (Mark stealing money), while others watched the videos in the reverse order. The participants were allowed to stop watching videos at any time — and they took advantage of that, turning off the reel when it began to run counter to whichever position they'd been assigned.

This turned out to be a smart move: the folks who avoided seeing disconfirming evidence, and therefore truly believed whatever they were arguing, wrote the most persuasive speeches.[1]

"We think it reduces the cognitive load, if you really believe what you're trying to persuade people about," Von Hippel says.

This is why defense attorneys generally don't ask their clients if they're guilty.

Deceiving yourself *about* yourself is even more powerful. In another study, Von Hippel and his colleagues found that middle schoolers who overestimate their athletic ability become more popular over the course of a semester than kids who make accurate self-assessments.[2]

There is, however, a point at which self-deception can backfire, Von Hippel notes.

"If I go through life believing that I'm 20 percent better than I really am—if I think I'm Bill-plus-20—I can get people to back down when they maybe shouldn't. I might be able to get women to go out with me who could do better. I might get guys to join my coalition. In all sorts of evolutionarily important ways, I can achieve goals that I wouldn't achieve otherwise," he says. "But if I believe I am Bill-plus-100, I'm probably going to get myself into trouble, because then I'll start pushing it too hard and I'll suffer for it. But as long as you don't overdo it, self-deception does make you more effective."

Face recognition is a crucial part of everyday social interactions, so believing that you're doing fine at it (even if you aren't) may be a particularly useful lie.

"If I've offended some people and I don't know it, then I can go back out there the next day with my head held high, oblivious to my own ineptitude," Von Hippel says.

Oblivious to my own ineptitude. Oh, how I miss those days.

"What am I supposed to do now that I know the truth?" I ask.

"You can't put the genie back in the bottle, but you can take advantage of the self-knowledge that you now have and try to improve yourself," he says.

OBLIVIOUS INEPTITUDE

"Didn't Socrates say that the most important thing is to 'know thyself'?" I point out.

"I'd say, 'Know thyself—but not too well,'" Von Hippel replies, chuckling.

I have no idea how to capitalize on my newfound self-knowledge—and it's really undermining my confidence at work. A faceblind reporter? It sounds like a punch line.

I know of a faceblind fellow with an even less likely career. So, to make myself feel better, I give him a call.

"If you're going to take someone with prosopagnosia and think of the worst job you could put them in, running a restaurant would sound like a good answer until you realize there's such a thing as being a politician, which is even worse," says John Hickenlooper, a US senator for Colorado. Before he got into politics, he ran a brewpub.

These careers would have been unimaginable to Hickenlooper back when he was a bespectacled, socially awkward kid. In elementary school, he was teased relentlessly, and he often kept to himself.

"I remember sitting, looking out my bedroom window—I was on the second floor—and looking over the hedge at the neighbors playing basketball in their driveway and...just dying to go over there. But I couldn't really remember who they were," he says.

"Me too!" I say, for the second or third time.

As he grew older, his social life gradually improved. In college, Hickenlooper wasn't exactly popular, but he did make "a pretty good sidekick," he says.

After college, Hickenlooper worked as a geologist for a petroleum company. "I would have been happy doing that for the rest of my career," he says, but when the commodities market crashed in the '80s, he was laid off and he decided to open one of the US' first brewpubs.

One day, while sitting at his own bar, Hickenlooper realized that his employees were much better at recognizing regulars than he was, and as a result, he sometimes "came off as a jerk." So Hickenlooper made a new rule for himself: if anyone so much as glanced his way, he'd shoot back a big, winning smile, or even give them a hug. This, plus some help from waiters who loudly greeted people by name, allowed Hickenlooper to become the social butterfly he was always meant to be.

"It's a good thing, in the restaurant business, to kind of treat everybody like they might be a friend," he says.

Even with this strategy, Hickenlooper failed to recognize his own brother when he made a surprise visit to the restaurant. This prompted a bartender to ask if Hickenlooper was faceblind. Hickenlooper had never heard of the disorder, but when he looked it up, it shed new light on a lifetime of social struggles.

"Understanding that prosopagnosia was a real thing was, at least for me, a huge confidence builder," he says.

That turned out to be the final boost Hickenlooper needed to become a professional schmoozer—which is to say, a politician. He loves campaigning as well as getting wonky about policy. But recognizing all ninety-nine of his colleagues is a challenge.

"Senators have a great deal of confidence and believe that they should be recognized immediately after one meeting, and that's just impossible," he says. "There are two really good senators—both of whom I really like, and now I can tell them apart. But it took me a year of really working at it... because they both have the same graying of their hair, and they both have broad foreheads and large eyes."

OBLIVIOUS INEPTITUDE

Though he was a lonely kid, Hickenlooper now sees faceblindness as more of a gift than an affliction, because it helped him learn to be extremely open and friendly.

The next famous faceblind person I call is Steve Wozniak—we FaceTime, actually, which seems appropriate for the cofounder of Apple.

"Woz" is not a developmental prosopagnosiac—he acquired prosopagnosia after a plane crash in 1971. He was unconscious for hours and experienced severe short-term memory problems for five weeks afterward. This near-death experience caused Woz to reevaluate his life and make some big changes.

"I got on the phone, called up Steve Jobs, and said, 'I'm going back to college, to finish my last year.'"

Woz managed to remain unnoticed by his fellow students—in part, because he used a fake name and changed his major from engineering to psychology. His crash left him wanting to learn about the inner workings of memory—and he was surprised to find how little we knew.

"We don't really know what's going on, or we would probably be able to find a way to fix it, put in a little circuit that has an alternate path," he says. "If we knew how the brain were wired, we could build a brain."

As I'm mulling that over, Woz says something surprising.

"Now, I was at a company, once, that *did* figure out how to make a brain."

"Oh yeah?" I reply, incredulous.

"Yeah, it takes nine months," he says, laughing.

I let out a relieved snort. "Good one."

I ask Woz if he worries about offending people by not recognizing them, and he says that's not been an issue at all. "When you're a celebrity, you meet so many people, no one expects you to remember them."

So there you have it, my fellow prosopagnosiacs: there's no cure for faceblindness, but you can alleviate the symptoms by becoming famous.

After chatting with Senator Hickenlooper and Woz, it occurs to me that prosopagnosia may have some upsides. Specifically, being faceblind has given me the ability to make people I've just met feel like we're already close personal friends. The ability to connect quickly with strangers is an excellent skill for a reporter.

I also suspect that I can thank my faceblindness, in part, for my sense of humor. After all, you can't take yourself too seriously when you're constantly making silly mistakes.

I trot out this theory with Paul Foot, a British comedian who is like an absurdist Eddie Izzard. Foot is faceblind, and he suspects that the disorder has also shaped his sense of humor. For instance, instead of hiding his bafflement, Foot has learned to lean into it to hilarious effect.

"My persona on stage is: I'm a little bit confused. Like I don't know what's going on," he says.

Prosopagnosia also gives you plenty of opportunities to deal with absurdity, and perhaps even to learn to enjoy it, Foot says.

"Years ago, I used to get embarrassed about not recognizing people, but now it's sort of fun—you've lost control of your golf cart and now you're enjoying the ride."

Talking with Foot makes me realize that prosopagnosia has given me loads of practice walking into situations where I don't know what's going on and trusting that if I pay close attention and ask the right questions, I'll figure it out. This is basically the job description for being a reporter.

Someone team me up with a super recognizer, and we'll be the next Woodward and Bernstein.

After my story comes out, I decide that it's time for me to stop bluffing and admit it when I don't recognize someone. This turns out to be a lot harder than it sounds.

My first chance at applying my new policy appears almost immediately. I'm taking the elevator up from the basement of the *Washington Post* building, where I've parked my bike. A tall, well-dressed woman gets on in the lobby.

"Hi, Sadie," she says.

She looks somewhat familiar—I think she works in the video department, but I'm not sure. "Sorry, can you remind me who you are? I'm literally faceblind," I say.

"Oh, don't worry about it—I'm always forgetting people's names," she says, before hopping off at her floor.

Maybe an elevator ride is not the best place to explain your neurodevelopmental condition.

The next person I run into is a young woman with curly brown hair who asks me how my birthday-party planning is going. She's obviously a friend, or at least an acquaintance, and I simply can't bring myself to admit that I don't recognize her.

"Great!" I reply. "I just rented fifty mermaid tails that you can actually swim in, and I ordered two remote-control balloons that look like sharks swimming in the air."

"Neat!" she says.

"You should come!" I say. "I'll send you the info."

"I have it," she says. "I'm bringing Gabe, if that's okay."

Who's Gabe? I wonder.

"That's perfect," I reply, excusing myself to attend to an imaginary deadline.

Shoot. I should have confessed. Then I could have asked a relevant question about *her* life. People must think I'm so self-centered.

At lunch that day, I meet my friend Lea at Union Station.

"Hey, Sadie, it's Lea," she says.

"I know," I reply, a little offended. I spotted her when she first walked through the doors — and I became 100 percent sure that this Lea-shaped creature was my friend when she smiled and beelined in my direction.

"Thanks, though," I add. It's sweet that people are trying to accommodate me, but it also feels weird. Is this something I want to lead with when I meet people? Will it eclipse other, more important aspects of who I am?

To disclose or not to disclose. This is a debate that comes up again and again (and again) in my new favorite hangout, Facebook's faceblindness support group — which I have come to think of as Faceblindbook. An increase in prosopagnosia awareness (thanks, in large part, to celebrity prosopagnosiac Brad Pitt) has made it easier to share and explain your problems. But it's still a big risk to bring it up, because some people won't believe you and others will think you're making a big fuss over nothing.

Thankfully, the burgeoning neurodiversity movement is helping to make these negative reactions less common. Neurodiversity activists argue that diagnoses like autism and ADHD should be seen as differences that often confer strengths as well as weaknesses. They are also pushing for more understanding and accommodation for people with atypical brains. It seems to me that this is in everyone's best interest. At a moment when humanity is facing existential-level challenges, we need all brains on deck.

There's already precedence for this kind of shift. Consider left-handed people. We used to think they were weird and possibly demonic; now we give them special scissors and we don't make a big deal about it. Kids who couldn't sit still in class were reprimanded and sent to the principal's office; now we hand them fidget spinners — or, even better, we give the entire class more time to run around outside. In the future, perhaps we won't be bothered if we see someone flapping their hands or doing some other kind of non-neurotypical behavior.

As for what accommodation might look like for faceblind people, I have one request: If you know someone's name, use it! Don't say, "Good point." Say, "Sophia here makes an excellent point, don't you think, James?" You will sound like a sleazy car salesman, but your faceblind and lethonomic friends will thank you.

Many faceblind people say they wouldn't want to be suddenly able to recognize everyone, that it would be overwhelming or would change some fundamental aspect of their identity, but I'm not among them. I'd love it if someone would give me augmented reality glasses with a face-identification module. Clearly there are privacy concerns around face identification. But this technology could be implemented ethically. For instance, everyone could have their own personal database of friends and acquaintances, consisting only of people who have explicitly given them permission to take their picture. And if you're worried about this technology getting loose, I have bad news for you: face-identification software is already commercially available and widely used — for instance, in home security systems.

That said, I don't really need much accommodation beyond simple understanding. Just try not to take it personally when I (or anyone) can't recall who you are. Like most faceblind adults, I've honed many effective strategies for dealing with my glitchy brain. Faceblind children, however, probably do need our help.

"Ben," age five, was having trouble adjusting to reception, the British equivalent of kindergarten. Usually bright and cheerful, his confidence drained away at the very sight of his school. He wasn't just shy. The little boy seemed to be *scared* of his classmates — all of them!

At his mom's prodding, Ben confessed that he was being bullied, and the teachers didn't believe him because he couldn't name the culprit.

"He literally couldn't pick out the kid who had pushed him on the playground, and the fact that this was a school with uniforms did not help," says Rachel Bennetts, senior lecturer at Brunel University London.

Concerned about her child's safety, Ben's mom did some Googling and found her way to Bennetts, who studies prosopagnosia in children. Ben turned out to have a rather severe case, and Bennetts worked with him to learn to recognize his mother — and her car.

"He tried to get into other moms' cars at school pickup," Bennetts says. "That's a major safety hazard, and it's not uncommon among kids with prosopagnosia."

This experience convinced Bennetts to switch her focus from basic science to treatment.

"That case really stuck with me, because a lot of researchers treat prosopagnosia only as a way to look at how face processing works. I know that's how I got involved in this area," she says. "But experiences like this made me realize that we're missing that human side of the story."

Sherryse Corrow came to a similar conclusion while studying

face recognition as a doctoral student at University of Minnesota, Twin Cities.

"We ended up getting contacted by this family with a kid who couldn't recognize faces, and we realized how isolating that experience can be," she recalls. "Word spread, and as more and more families all over the country began reaching out to us, we realized that there was a need and an opportunity here."

In 2012, Corrow and her colleagues organized ProsoKids Weekend, a mini-conference for faceblind kids and their parents. While the adults attended talks by face-recognition scientists, the kids went on field trips, did arts and crafts, and brainstormed ways to work around their disability.

"It was a delicate situation, because some of the kids' parents didn't want to label their kids, so we didn't tell them that they were faceblind. But when we asked them to talk about things they were good at and not so good at, kids who hadn't said anything to their parents about face recognition would be like 'I don't know who people are,'" Corrow recalls. "Some parents were really surprised."

Some parents were worried that if they told their children about their diagnosis, it would undermine their social confidence. But prosopagnosic kids already tend to be socially anxious, and arming them with self-knowledge seems to help them cope, Corrow says.

"It was freeing for them to learn that they could get themselves out of awkward situations—for instance, by asking their close friends to identify other kids. Or, if they realize *I don't know who I'm playing with right now,* they could just say, 'Who are you, again?'"

Due to a lack of funding, ProsoKids Weekend was never repeated, and there's still not much research aimed at helping people with prosopagnosia. But the tide may be turning. Bennetts and her colleagues, for instance, are helping children sharpen their face-perception skills, using the commercially available board game Guess Who? The game begins with both players picking a face card out of twenty-four possibilities. The goal is to guess which face the other player has, and you do it by asking yes or no questions to narrow down the field. So, in the original game, you can knock down half of the cards simply by asking if the face is male or female.

In the original game, the characters have different hair colors, hats, glasses, and other external characteristics. Bennetts replaced those cards with faces that were all very similar looking, to make players pay attention to facial details, like the distance between the eyes or the size of a character's nose.

Bennetts and her colleagues found that children aged four to eleven who played her version of the game improved their face-recognition skills 7.5 percent more than a control group who played the original game.[3]

"All these children were neurotypical, because faceblind kids are so hard to find," Bennetts says. "But the kids that were below average to begin with showed the biggest gains. So we're hopeful that it will be even more effective for faceblind kids."

While shoring up kids' weaknesses is a noble goal, it might be more effective to build on their strengths, says Judith Lowes, a psychology PhD student at the University of Stirling in Scotland.

"Instead of trying to make people with atypical face recognition or face perception more like typical perceivers, it might be better to help them develop coping mechanisms, like using hair and voices," she says.

This tailored approach won't necessarily contribute to the science of face perception, but it could make a big difference in people's lives.

"Children with prosopagnosia face a lot of challenges," she says. "Even just a small improvement in practical person recognition might help."

While we have a pretty good idea how acquired prosopagnosia happens (damage to your right hemisphere's FFA), the origins of developmental faceblindness remain unknown. Research by Ingo Kennerknecht suggests that it's caused by a single gene.[4] "This is a clear autosomal dominant disorder," he says.

Brad Duchaine agrees that there's a genetic component, but he suspects there are usually multiple genes involved. That's because, while faceblindness does run in families, it tends to get diluted the further away you get from "patient zero." So, if a mom is totally faceblind, she might have kids who are just a little below average, and grandkids who are normal.

"If it was just one dominant gene, it would kind of just go on and on" within one family, Duchaine says. "But we think there are many different genes involved, and the individual with DP [developmental prosopagnosia] just happened to get a bad shake of the dice. And you revert to the mean, the more you move away from the individual."

Of course, both researchers could be right: there may be a version of faceblindness that's caused by just one gene and another variant that's caused by many. But neither of those stories explains me. I haven't been able to find a single family member who has any degree of prosopagnosia.

So what happened to me? Was I dropped on my head as a baby?

When I ask my dad this, over dinner one night, he bursts into tears. (Cliff-hanger!)

7

FEAR, UNMASKED

On a bright, humid morning in the summer of 1996, I walked to school while reading *The Hitchhiker's Guide to the Galaxy*. My favorite character, Zaphod Beeblebrox, had run afoul of the Vogon, a race of highly bureaucratic aliens. They were about to lock Zaphod in a torture chamber when the world turned sideways. My book flew out of my hands as I folded over the hood of a car.

Had I been hit? No, I had walked into a stationary car — one that someone had, very rudely, parked in the middle of their own personal driveway. I scooped up my belongings and rushed away.

When I got to school, I slipped into the classroom without attracting the attention of the driver's ed instructor, a ruddy-faced man with a Foghorn Leghorn accent. He was delivering good news.

"Ay-as I was saying, yuh all passed," he said.

An assistant began passing out our learner's permits.

DO I KNOW YOU?

The summer school driver's ed test was much easier than the DMV version, which I'd failed twice. I still didn't know the meaning of the various kinds of road stripes, and I wasn't totally sure what side of the road was "right," in either sense of the word. But as far as the state of Florida was concerned, I was ready to roll.

Mr. Leghorn explained that we'd be divided into two groups—half of us would stay inside, while the other half went out driving—and we'd switch places tomorrow. Everyone besides me was dying to go driving, so it was easy to land myself in the "stay inside" group. I'd already had enough excitement for one day, having been both the victim and the perpetrator of a hit-and-run.

Once the classroom emptied out, an assistant coach turned on the TV and it flickered to a scene from *Jurassic Park*—a movie that, for some reason, played constantly on my high school's closed-circuit system. Many of my classmates could recite the whole thing by heart, while the more advanced students pointed out continuity errors.

"See how the eggs are on dirt?" said a freckle-faced boy. "Now it's sand."

I pinned *The Hitchhiker's Guide* open with my elbows and read by the flickering light of the TV. Laura Dern plunged her arms into a mountain of poop. A tiny dinosaur blinded a treacherous computer programmer. Zaphod emerged from the torture chamber, miraculously unscathed.

The next day, when the coach sent my half of the class out to drive, I hung back. If anyone noticed, I'd pretend not to have heard the announcement. Being a known space cadet has its advantages.

No one noticed. Not that day, or the next one, or the one after that.

For six weeks, I sat in a trailer in my high school's parking lot, working my way through Douglas Adams's entire oeuvre.

I knew skipping out on the driving part of driver's ed was not the best idea, but it took me a while to figure out why. The night before the driver's test, I suddenly realized the impossible predicament I'd created for myself. If I refused to take the driver's test, I would get an F on my actual report card—tanking my GPA and sentencing me to a mediocre college and, by extension, a mediocre life.

But if I attempted the test, having never driven before, I'd probably still fail and possibly also die.

There was no other option. I had to pass the driver's test. *You're smart*, I told myself. *You'll figure it out.*

The next day, the coach took us out to the school parking lot and explained that we'd be taking the test there. I was only slightly relieved. A closed course is certainly easier and safer than the open road, but it also meant that my classmates would have front-row seats to my (probably) spectacular failure.

Mr. Leghorn explained that we would be circulating through seven stations—at each one, we'd demonstrate a required skill. I was assigned to start at parallel parking, but that sounded much too hard, so I wandered over to "quick stop," instead.

An assistant coach demonstrated the skill for us: Put on your seat belt, check your mirrors, drive in a straight line, and then stop. Quickly.

I hung back and watched many, many kids go first. Everyone passed, except the freckle-faced boy, who forgot to put on his seat belt. (They reprimanded him and then let him try again.)

Finally, it was my turn. I casually approached the car, and then—mimicking my classmates—I made a big show of fastening my seat belt and adjusting the mirrors. *Okay, there are only two pedals*, I told myself. *All I have to do is figure out which is which...*

"Go," said the assistant coach. I shifted the car into drive, and we drifted forward.

"Stop," he called out. I tentatively pressed the big pedal, and the car stopped.

"Good job," the coach said. "That was your last station, right?"

"Right," I replied, passing him my form.

He scribbled on the paper, handed it back to me, and walked away.

I stared slack-jawed at the form. The coach had signed the line indicating that I'd passed the test. Not just that one station — the whole test.

I passed! I PASSED! How the *hell* had I passed?

Later that week, my dad drove me to the DMV and I traded that piece of paper for a driver's license — one that I have been renewing ever since.

"Would you like to be an organ donor?" the DMV lady asked.

"Sure," I said.

It was, I figured, the least I could do.

Even though my faceblindness story came out months ago, there are still little piles of studies all over my apartment. Every time I try to gather them up and throw them out, something gets in the way. I can't find the key to the dumpster room. Dishes need washing. I remember a call I need to return.

This is not like me. I love throwing things away. Half-full cans of seltzer, unopened mail, tiny electronics that Steve scatters like fairy dust — into the garbage they go! And yet, I can't seem to toss these damn studies. I eventually shove them into a shopping bag and banish them into the back of a closet.

Bigger problems emerge. The *Washington Post Express* folds

and I'm laid off. The pandemic starts and DC goes into lockdown. Steve and I and the cats start spending *a lot* of time together.

The bright side? Steve can work from home, and—as a newly minted freelancer—I can too. Theoretically. The assignments are not exactly rolling in, and Ancestry.com knows a soft target when it sees one. Their ads follow me from Facebook to Amazon to YouTube. *Discover what makes you uniquely you. Bring your backstory to life.*

Steve's days are spent mostly in Zoom meetings. I thought I'd married a soft-spoken guy, but this turns out not to be the case when he's at work. Just as I'm on the cusp of figuring out some family mystery, he says something in a voice loud enough to keep the International Space Station in the loop.

"WE NEED TO CONSIDER THE IMPACT OF THE THERMODYNAMIC OSCILLATION ON OUR HYPERPARAMETER MATRIX," Steve says. "INTEGRATING A RECURSIVE AUTOENCODER WILL MINIMIZE DATA DISCREPANCIES."

This isn't even worth eavesdropping on.

I love Steve, but we were never meant to spend every second of every day in an 820-square-foot apartment. Even the cats are feeling the effects of too much family togetherness. They try to escape every time I open the door.

Desperate for sunlight and a glimpse of the sky, I relocate to our tiny "Juliet" balcony. I never used it before the pandemic, but once I string up a hammock, it quickly becomes my favorite place. I start spending all my time in my "office," "working"—which mostly entails calling up elderly family members and shaking them down for details on deceased relatives.

None of them sound the least bit faceblind. There's a silk magnate, a famous rabbi, and some card sharps and bootleggers—all jobs that require good people skills. Plus, all of my living relatives are scoring average or above average on the face memory test.

So what the hell happened to me?

I feel a sudden need to revisit my research. Unfortunately, I am not wearing pants. To avoid making a surprise appearance on Steve's Zoom call, I crawl through the bedroom to the closet, grab my bag of studies, and crawl back out to the balcony.

I'm looking for one paper in particular—a study on environmental causes of prosopagnosia, instead of the usual genetic culprits. I spot the paper quickly, as it's covered in highlighter.

In 2001, researchers at McMaster University in Canada recruited fourteen children who were born with cataracts in both of their eyes and had them removed when they were between the ages of two months and one year. About nine years after their surgeries, these kids showed prosopagnosia-like symptoms. While they were good at identifying individual features, they had trouble taking in the face holistically—and as a result, had trouble distinguishing between similar-looking faces.[1]

This leads me to a follow-up study. This time, the kids had cataracts in only one of their eyes—and the kids with left-eye cataracts ended up with face-processing deficits, while the kids with right-eye cataracts processed faces just fine.[2]

That means early exposure to faces through your left eye is critical to developing face-recognition skills. The right eye? Not so much.

This finding makes sense because, in infants, the left eye connects mostly to the right hemisphere*—so if the left eye is cloudy,

---

*   In adults, both eyes project to both hemispheres.

that means the right FFA (which is the one that does the heavy lifting when it comes to face recognition) is not getting the visual input it needs to develop.

Reading this is so exciting, I nearly tip out of my hammock. *My* left eye/right FFA missed out on seeing faces when I was a baby! I didn't have cataracts—I had a condition colloquially known as lazy eye, which caused my brain to ignore information coming from my left eye. To make ignoring that information easier, my brain pointed my left eye at the blank screen that is the side of my nose.

I feel like a detective who has just found a key piece of evidence on a cold case.

A trail of clues leads me to a 2015 study that directly supports my theory: people with left-eye amblyopia (and not right-eye amblyopia!) have lifelong face-processing deficits.[3]

I hit PRINT and run inside to snatch up this study. There's definitely more to the mystery of my faceblindness—after all, not everyone with lazy eye is faceblind—but I feel like I've found at least one of the smoking guns.

Amblyopia, strabismus, and stereoblindness: these terms all overlap, while also meaning slightly different things.

Amblyopia is when your brain suppresses information from one of your eyes. Strabismus refers to eyes that are misaligned. Amblyopia and strabismus frequently co-occur—and they can create feedback loops that cause both conditions to get worse. Babies who have these problems grow up using only one eye at a time, and therefore cannot see depth. They are stereoblind—and that's the word I'll be using most of the time, because it describes the functional problem and makes more intuitive sense.

Like many amblyopes, I thought my condition had been cured by eye-realignment surgery when I was a kid. But when the doctors yanked my eyes into alignment, it had no effect on my vision. For me, these surgeries amounted to braces for eyeballs. I looked normal to other people, but I remained quite stereoblind.

My experience is far from unique. More than fifty thousand eye-realignment surgeries are performed on children (including babies!) every year,[4] and fewer than 0.5 percent result in normal stereoacuity[5]—which is the ability to detect how far away objects are in the absence of other cues. In many cases, eye-realignment surgery makes a child's eye problems invisible to other people, but the vision impairment remains.*

I thought my vision was totally normal—until, in 2009, a Harvard neuroscientist named Margaret Livingstone suggested otherwise.

This happened during my stint as a science writer for the American Psychological Association. I was working on a story about Livingstone's research, including a study in which she found that art students are more likely than students with other majors to have some degree of stereoblindness. She also compared portraits of famous artists to portraits of famous politicians, and she found that the artists were more likely to have wonky eyes—which meant they were, most likely, stereoblind.[6]

Her conclusion? If you live in a flat world, you will find it a little easier to render our 3D world on a 2D canvas. But I was more focused on something she'd said before—that having misaligned eyes causes you to see the world differently than other people do.

---

* I am not a doctor, bodies are complicated, and you and/or your child may well benefit from this surgery. Please don't take my personal story as medical advice or write me angry letters. I fully support whatever medical decisions you and your healthcare professionals decide are right for you.

"But *my* eyes are misaligned," I said. "And I'm not stereoblind."

"Are you sure about that?" she replied.

The next week, I happened to have a routine eye exam.*

"Just so you know, I can be a little flinchy," I told the nurse, as she approached me with an eyedropper. "I had a lot of eye problems as a kid..."

"I'm sure you're not that bad," she said offhandedly.

I am that bad. Ten minutes later, we were both soaking wet, reinforcements had been summoned, and there was talk of locating something called a lid speculum.

As I attempted to eye-drop myself, the nurse scribbled a note in my file—probably a warning to other eye-care professionals.

The doctor arrived—and from the casual way he approached me, I knew he hadn't read my notes.

He shined a bright light directly into my eyes as I clenched the arms of the chair while holding my eyes (and, for some reason, my mouth) wide open. Then, he told me to put my chin on a little shelf. *Oh no*, I thought, *it's time for the eye puff test.*

"Just so you know..." I started to say. The puff stopped me midsentence. I yelped and involuntarily kicked the floor, causing my chair to fly backward into the wall.

The doctor and I stared at each other, wide-eyed, in silence. Then a thin, animal-like wail echoed through the office. My freak-out had upset a fellow patient, a small child by the sound of it.

"Sorry," I said, though I didn't really mean it. That kid was going to find out about ophthalmologists soon enough. All I did was give him a little advance warning.

---

* Because of my complicated eye history, I have to see an ophthalmologist every few years. Ophthalmologists are medical doctors, while optometrists have a different kind of degree.

"Almost done," the doctor said triumphantly. Then he handed me special glasses and a puffy plastic folder with pictures of bull's-eyes in sets of four. My job was to select the one that seemed 3D.

"None of them pop out," I replied. "I always fail this test. I can't see the fly either."

"Oh, right," he said, glancing down at my records. "You're stereoblind."

Stereoblind. I'd never heard this word before, and now it was coming up all the time.

"But I'm a terrible artist," I said.

"I bet you're not very good at tennis either," the doctor said.

"I'm the worst!" I exclaimed, delighted. "My parents sent me to tennis camp for years, and I still can't hit my own serve toss."

He explained that while most people combine their two fields of vision into a single image, I switch back and forth between my left eye and my right eye. As a result, my depth perception sucks.

Three decades of missed Frisbees and spilled drinks suddenly made a lot more sense. I called my dad the moment I got back to work.

"Hey, remember how bad I was at softball? How I never hit the ball the entire season?"

"Mmmmhmmm," my dad said. He sounded suspicious— which was fair. I didn't usually call him in the middle of the day to talk about sports.

"Well, it's not my fault that I couldn't hit or catch a ball. I just went to the doctor, and he said that I'm stereoblind! I can only see out of one eye at a time!"

My dad laughed. "Are you sure that you're not just a klutz?"

"I think this is *why* I am a klutz," I replied.

We started talking about the surgeries we thought had solved my eye problems. The first one had been, apparently, rather traumatic for my dad.

"I still remember them wheeling you off," my dad said, tears in his voice. "My tiny little baby."

I was one, so I don't remember my first surgery. I do remember my second surgery, though, at age nine. I was scared. Everyone kept telling me that the surgery was safe, but I didn't believe them. Then, just a few days before I was supposed to go under the knife, something happened that confirmed my worst fears.

My mom and I were at the mall, and I idly picked up a music box featuring a carousel horse. "Would you like that?" my mom asked. Without even waiting for me to answer, she bought it for me.

This felt highly suspicious, so a little later, I picked up a stuffed animal, a floppy golden retriever puppy. My mom bought that for me too.

Welp, that cinches it, I thought. *I am definitely going to die.*

(I didn't die. Eye-realignment surgery is quite safe.[7])

"Can you get your eyes fixed now?" my dad asked.

"Nah," I said. "But it's no big deal."

My eye doctor, and many eye doctors thereafter, told me that the critical period for learning to see in 3D had long since passed—but it didn't really matter, because I got by just fine.

Case closed. Or so I thought. Because more than a decade later, I discover that my ophthalmologist was wrong on both counts.

1. Being stereoblind *is* a big deal—and it has subtly affected nearly every aspect of my life.

2. Stereoblind adults *can* learn to see in 3D—and the few who have done this say it transformed their lives and even changed how they *think*.

Move over, prosopagnosia. There's a new sheriff in town.

If you're like most people, your brain automatically combines information from each of your eyes into a single, seamless, 3D scene. This is an impressive trick, and scientists are far from understanding how it happens. What we do know is that it involves some complex calculations and a few educated guesses—all of which you learned to do before you were four months old, by trying to grab everything in sight.

As I mentioned before, some infants aren't able to get their two eyes focused on the same object. To avoid seeing double, their brains suppress information coming from one of their eyes. Often, the brain will intentionally pull the weaker eye farther out of alignment, to make that conflicting information easier to ignore.

I knew some of this, but I never put it all together until I read *Fixing My Gaze*, by neuroscientist Susan R. Barry.

The book tells an oddly familiar story. Despite three childhood surgeries that cosmetically lined up her eyes, Barry was stereoblind her entire life. It hasn't exactly held her back—she's a Mount Holyoke College neuroscience professor and a bestselling author. And yet, it did prevent her from pursuing a career in ornithology. She always wondered why everyone else was so much better at spotting birds than she was.

When Barry hit her forties, she noticed that her vision was becoming somewhat jumpy and unstable, so she signed herself up for vision therapy. Along the way, something unexpected happened, something that most doctors and neuroscientists (including Barry) thought was impossible. She started seeing in glorious, vivid 3D.

With this new way of seeing the world, Barry found that she became better at spotting birds, and she also became more confident about driving—something that had always made her anxious.

When I read this, the bit about Barry being afraid to drive struck a chord. Is that a stereoblindness thing?

As it turns out, it is. Only 50 percent of stereoblind eighteen-year-olds have a driver's license, as compared to more than 70 percent of teens with normal vision.[8] Driving scares them, and that fear persists into adulthood. Many stereoblind adults report high anxiety while driving, especially at night.[9]

If you'd asked me, before reading Barry's book, why I never learned to drive, I would have lectured you about fossil fuels or waxed poetic about the joys of public transportation. Suddenly, I see what was probably obvious to everyone else: I never learned to drive because I was afraid.

My insistence that I didn't *want* to drive kept me from thinking about all the things I was missing out on. Now that that's gone, it opens floodgates of regret.

I grew up in Florida, and I almost never went to the beach, because it was too far to walk. I went to college in gorgeous western Massachusetts, and I never left campus to explore the nearby towns. Every summer in DC, I dream of hiking in the mountains and finding secret swimming holes. But I can't get myself there, so I end up kayaking up and down the Anacostia, trying not to think about all the raw sewage I'm paddling through.

Suddenly, I'm furious at myself for letting fear rule my life. It's unacceptable. It has to stop now—like, *this very instant*. I sit up in my hammock and holler through the bedroom window.

"Steve!"

No answer.

"Steve!"

No reply.

I get up and find Steve in bed. To get his attention, I block his view of his computer screen by draping myself across his chest.

"I want to learn to drive," I announce.

Steve seems suddenly alert. He's thinking about me driving, and what it means for his personal car.

"Is that a good idea, with your vision problems?" he says.

"Yes, obviously," I say. "Don't be rude."

Actually, the research is a little mixed.[10] One study found that stereo-deficient taxi drivers tend to get into more crashes than drivers with normal vision, and that stereo-deficient truck drivers get into more severe accidents than their 3D-seeing counterparts. Another study found that stereoblind people aren't very good at driving through slalom courses but that we do fine with more typical driving challenges. Among noncommercial drivers, people with poor 3D vision do not get into crashes any more often than regular folks do. This is why all US states and most countries require only one working eye to get a (noncommercial) license. No need to tell Steve all the details, though. I should be capable of driving as well (or maybe as badly) as everyone else.

Later that week, I catch Steve browsing Redfin. I guess I'm not the only person in this apartment who could use a little more space.

"We could actually *save* money by moving to a bigger place, out in the country. And it would be a lot easier for you to learn to drive," he says. Then the kicker: "I think the cats would be happier too."

FEAR, UNMASKED

I've always considered myself a city person, but now I'm not so sure. What if that's just another smoke screen, keeping me from seeing how much my fear of driving has limited my life? What if I belong in the mountains, hiking and kayaking and nursing injured baby animals back to health?

What if, and I know this is a long shot, but what if a dramatic change in scenery will help me learn to see the world in an entirely new way, in glorious 3D?

I know one way to find out.

"Okay," I say. "Let's do it."

8

STUDENT DRIVER

The moment I lay eyes on the 1988 Ford F-150, I fall in love. We have so much in common. She's roughly my age, she's caked in dirt, and her cab is tiny compared to her big bouncy butt—I mean, bed.

I can already imagine all the fun we're going to have together—exploring the mountains, going to fiddle festivals, finding swimming holes...

"She's perfect! Let's get her."

"Her?" Steve says.

"Yeah. Check out those hips."

My plan had been to work on my vision and *then* learn to drive. But circumstances pushed our move date from "the vague future" to "next month." The details are too tedious to relate here, but let me just say, there's no such thing as a "practice bid."

I'm not complaining, though; my new house is gorgeous—a modern cabin in the woods, with big windows and porches on three sides. The downside to being in the middle of nowhere is that there are no restaurants. There's a pizza place that supposedly delivers, but the teens who work there leave the phone off the hook to discourage orders.

I need to buy a vehicle—and fast. Otherwise I'm going to have to learn how to cook.

A man with windblown white hair ambles over and slings an arm over the side of the truck.

"Hi, I'm Frank," he says. "You here to see the Ford?"

"Yes!" I exclaim, immediately failing to play it cool.

"The ad said that the truck has only 43,000 miles on it," Steve says. "How is that possible?"

"That's what the odometer says," the salesman replies with a shrug.

He opens the door, and we see that the odometer reads 43,203. However, there are only five available digits.

Steve looks skeptical, but I remain enthusiastic.

"Can we take it on a test drive?" I ask.

Frank explains that he's sent his son-in-law to fetch some gas, so we have to wait for him to get back. It seems odd, a used-car lot without any fuel on hand, but we don't have a whole lot else to do, so we wait.

"Is this your lot?" I ask.

"Yup," says Frank. "I'm also a pastor. You come up Route 11? I preach at the church by the U-Haul."

Frank has a preacher's gift of gab. Over the next hour or so, he tells me all sorts of things, including how his wife recently recovered from COVID—thanks, somehow, to West Virginia governor Jim Justice. This is quickly followed by information on Frank's own medical woes.

"The hospitals are so backed up right now, they canceled my cataract surgery—again!"

"Oh no!" I say.

"I'm nearly blind," Frank adds. He holds out his hand and wriggles his fingers. "Can't hardly see my own hand in front of my face."

Abruptly changing the topic, Frank starts telling me how he inherited this car lot from his father. The old man had an unfortunate habit of buying broken-down classic cars and forgetting where he parked them. Over the years, the cars have rusted and a forest has grown up around them. And this is why Frank recently found an old Corvette with a deer leg sticking out of the hood.

"It must've jumped on the car and punched through the metal—got stuck and chewed his own leg off," Frank says.

As an animal lover, I'm horrified; but as a writer, I'm thrilled. I'm imagining the muscle car ensnared in a net of vines, slowly sinking into the wet woods. Bright green moss spills across leather seats, ferns unfurl in empty wheel wells, and jutting out of the hood like a radio antenna is the desiccated leg of an unlucky ungulate.

What else is in Frank's woods? Before I get up the nerve to ask for a tour, his son-in-law returns with the gas.

"Test-drive time!" I announce.

"Actually, my insurance doesn't allow that," Frank explains. "I'll have to drive, and you two can squeeze in."

I peek inside the cab and consider our options.

"Well, there are three seat belts, so I guess it's okay," I say.

I'm congratulating myself on my safety-mindedness when I suddenly remember two key facts: Frank's wife just recovered from COVID. And until he gets his cataracts removed, Frank is currently blind.

Maybe he was just exaggerating.

"Is anyone coming?" Frank asks.

The road is obviously empty for miles in both directions.

This doesn't seem very safe, but bailing, at this point, feels overdramatic. And heck—for all I know, being chauffeured by a vision-impaired preacher is standard truck-buying procedure.

"You're good to go," I say.

We clatter down the road, a mile and back, without incident. The shocks on this thing are not very effective—I can feel every individual piece of gravel that we roll over. But other than that, the truck seems to run well.

"We'll take it," Steve says.

We sign a bunch of papers, hand over $2,300 in cash, and make arrangements to get the vehicle delivered.

"You don't want to drive it home?" Frank asks.

"Nah, I don't know how to drive yet," I explain.

Frank's bushy white eyebrows jump a little.

"You picked a tough truck to learn on," he says.

"I'm a tough lady," I say, with false bravado. What else can I say? We'd already bought the thing.

On the way home, we get stuck behind a slow-moving tractor piloted by a teenage boy with a goat riding shotgun. West Virginia, it occurs to me, is a lot like my home state of Florida: colorful, lawless, a little anarchic—a good place for someone with a questionable driver's license to take to the road.

"I'm feeling better about this whole learning-to-drive thing," I tell Steve. "I mean, if a blind guy can do it, how hard can it be?"

My truck arrives early the next morning, and I run outside, barefoot and in my pj's, to greet it. After snapping a selfie, I scurry back indoors and wake up Steve.

STUDENT DRIVER

"Can we go driving now?" I ask.

He checks the time and leaps out of bed. "I have a conference call in ten minutes," Steve says.

That call runs into another, and there's another one after that. Poor Steve, trapped in Zoom's nightmarish panopticon. He's probably going to lose his voice. At about noon, I get tired of waiting and decide to take the truck out by myself. After all, I have a valid driver's license and a perfect driving record. In over forty years, I've never gotten a single ticket.

"I'm just going to go around the block," I text Steve.

"Watch out for the ditch by the driveway," he texts back.

The only thing I recall from driver's ed is that one should never, ever wear sandals while driving. So I trade my flip-flops for Crocs and then climb into the truck. It smells like mold and old cigarettes. The impact of my rump on the seat sends up a plume of dust.

I reach underneath the seat for a lever that might slide me forward. Instead, my hand brushes something soft. Tiny claws grapple for purchase on the back of my hand, and a small gray creature flings itself out of my still-open door.

This is not the kind of wildlife encounter I'd been dreaming about.

I insert the key and try to turn it, and it refuses to budge. I'm forgetting something, but what?

"Hey, Siri," I say, feeling sheepish. "How do you start a car?"

Siri sends me to a website with step-by-step directions. Oh yeah! Now I remember. You push the key in while rotating it. I try again. The key moves, but the truck still won't start. I go back to the internet and get another idea. This time I perform the key-pushing-turning maneuver while simultaneously pumping the gas.

The truck grumbles reluctantly to life.

"Wooooo!" I shout. "Attagirl!"

I put the truck into reverse, back out of the driveway, and glide smoothly onto the street. Feeling like a goddamn pro, I shift into drive and begin plodding slowly around the block.

I consider trundling all the way to the Dollar General for a celebratory Diet Coke. But that would mean attaining speeds upwards of twenty miles per hour, so I play it safe and stick to the one-block plan.

As I'm pulling triumphantly into our driveway, the truck abruptly tilts forward and then refuses to budge. This must be the gully Steve had been talking about.

It takes a few hours, but he eventually notices my unconventional parking job.

"I told you to watch out for the ditch," Steve says.

"I *did* watch out for it," I say. "I just didn't *see* it."

I've been reading obsessively about stereovision, so I know exactly why Steve can see a depression in the ground that's invisible to me. It's the same reason he can spot birds in trees, catch Frisbees, park in tight spaces, and navigate our new house without constantly stubbing his toes.

My jealousy is so sour I can almost taste it.

In the waning days of October 1956, Soviet-occupied Hungary experienced a moment of freedom. Students took over the radio station in Budapest, and protesters in other cities followed suit. They broadcasted their demands: an end to the post–World War II Soviet occupation of Hungary, freedom of speech, freedom of the press, a withdrawal from the Warsaw Pact, and free elections.

At first, the Soviets blinked. The USSR began to move Red Army troops out of the capital, which sparked a series of bold reforms. A new prime minister was appointed, the secret police was disbanded, and groups of Hungarians known as revolutionary councils began meeting to discuss the nuts and bolts of local self-rule.

This joyful interlude came to a sudden end just twelve days later, on November 4. A thousand Soviet tanks and thirty thousand soldiers descended on Budapest. Bridges were bombed to keep people from fleeing. Hungarian civilians fought Russian soldiers in the streets.

A young civil engineer named Bela Julesz did not join the melee. Instead, in the dead of night, he and his wife snuck out of their apartment and stole through the deserted streets, making their way to a tributary of the Danube. The two swam to the other side, to Austria, and then hiked to a radio station.

Back in Budapest, eleven other couples gathered in the Juleszes' apartment, waiting anxiously by the radio to hear whether the two had made it. They would know that the escape route was safe if the radio announcer said the code, "The teddy bear is tucked up in bed."

What the announcer said instead was "The teddy bear is tucked up in bed, and his glasses are on the nightstand."

The room fell silent. What was Bela trying to say? Had they been captured? Was the escape route compromised?

Suddenly, Bela's father clapped himself on the forehead. "His glasses! They must still be in his bedroom," he said.

All twelve couples and Bela's glasses made it to the West.[1]

The Juleszes landed in a New Jersey refugee camp, where Bela asked to speak with someone from Bell Labs, a world-famous telecommunications research institute. The lab sent mathematician Sergei Schelkunoff—and Julesz asked him if he was the same Sergei Schelkunoff who had authored a book on microwave antennas. Julesz explained that, while working on his second PhD, he'd translated Schelkunoff's book into Hungarian. Julesz had no proof of any of this, but luckily he had an amazing memory, so he was able to recite the book's introduction verbatim.

Floored by this performance, Schelkunoff whisked Julesz to Bell Labs, where the Hungarian ended up giving an extemporaneous lecture on advanced techniques for encoding TV signals. Even though Julesz was from a Warsaw Pact country and sounded a little like Dracula, he was clearly a genius, so Bell Labs snapped him right up.

This turned out to be a great stroke of luck for both parties. In the mid-twentieth century, Bell Labs was a playground for brilliant engineers and scientists, all of whom were given access to the most advanced technology and the freedom to pursue any line of inquiry they pleased. This combination proved to be quite potent. Before the breakup of Ma Bell in the 1980s, Bell Labs churned out eleven Nobel laureates, more than twenty-six thousand patents, and many major inventions—the transistor, sonar, lasers, satellite arrays, and the UNIX operating system, to name a few.

Julesz's first assignment was to reduce the bandwidth requirements for television broadcasting. To figure out how to trick the human visual system into thinking it's seeing more than it is, he read up on vision science and quickly spotted a major flaw in a leading theory.

In the 1950s, most scientists believed that the brain calculated the distance from a given object in much the same way a surveyor would, through triangulation. It was a plausible theory, but Julesz instantly knew it was false, because to triangulate something, you must first identify that object as a distinct thing.

From his experiences in World War II, Julesz knew that

humans see in 3D even when we have no idea what we're looking at. He knew this because, during the war, spy planes would take two photos in quick succession—capturing a bit of suspicious land from two slightly different spots in the sky. These photos were developed and then placed into a stereoscope—a device that shows one picture to each eye. (You might have encountered a toy version of this contraption in the '80s, called the View-Master.) When people with good stereovision viewed pairs of aerial photographs through the stereoscope, their brains combined the two images into a single, 3D scene. This gave them the perspective of a giant with eyes about a plane-length apart. Concealed objects—for instance, a tank covered in hay, sitting in a hayfield—became glaringly obvious.

If stereovision didn't depend on triangulation, what was the alternative? The brain would have to make a pixel-by-pixel comparison between the images that fall on each of the retinas. That would involve making sure that every neuron relaying information from a particular point in the right eye sat next to the neuron connected to the corresponding pixel in the left eye—a daunting task, given that each human retina has 100 million neurons. This struck many scientists as impossible. They argued that spy-plane photo interpreters must use faint, perhaps undetectable features in the images to triangulate depth.

To convince the naysayers, Julesz created two images that were pure noise when seen one at a time, but when they were viewed through a stereoscope, a ghostly shape would emerge—an image that's now known as the random-dot stereogram (RDS).

Julesz's invention elegantly debunked the reigning theory of stereopsis. It was a major achievement, and it brought him international acclaim. In the following decades, Julesz built on this foundation and won many awards, including one of the first

MacArthur "genius" fellowships. But none of this dulled the sting of betrayal when, in 1979, Julesz's former student improved on his design and eclipsed his fame.

Christopher Tyler was lying on his bed in his Sausalito houseboat, thinking, *Wouldn't it be cool if you could see a stereogram without the glasses?* Suddenly, it occurred to him that this was possible. All he had to do was hide the 3D information in a repeating pattern, and people's brains would do the rest.

If you lived through the '90s, you should be getting flashbacks right about now. Tyler's invention, the autostereogram, sparked a major pop culture phenomenon: Magic Eye puzzles. These densely patterned, vaguely psychedelic images were everywhere—on T-shirts, cards, coffee mugs, tote bags, calendars, and screen savers.* They were especially ubiquitous in bookstores, where they'd be surrounded by strangers giving one another advice. "Focus on a point *behind* the picture." "Blink." "Relax."

I and millions of other stereoblind children got headaches trying, and failing, to see these things. It was a cruel blow, especially since many of us had only just recovered from our failures with the View-Master.

I call Tyler to give him a piece of my mind.

"You really made the '90s difficult for me," I say. "I felt so inadequate."

"I'm so sorry," he says in a crisp British accent. "But you were in good company—half of the population couldn't do it."

Tyler invented the process behind the Magic Eye puzzles, but he didn't make any money from the craze. The people who did were computer programmer Tom Baccei and artist Cheri Smith. They transformed Tyler's field of black and white dots into

---

* If you don't know what a screen saver is, ask a millennial.

brightly colored, wallpaper-like patterns, and branded their creation Magic Eye.

"I would go into shopping malls and the bookstores would have all these autostereogram posters," Tyler recalls, "and people would come up to me and say, 'You gotta see this. This is so great.'"

Tyler kept a low profile, but occasionally culture reporters would hunt him down. He was always careful to explain that Julesz's invention, the random-dot stereogram, was the real breakthrough. Unfortunately, that nuance often didn't make it into the papers, and Julesz became convinced that Tyler was taking credit for his idea.*

"He just felt very bad about it," Tyler says. "Our friendship never recovered from that."

Indeed, Julesz's random-dot stereogram quickly became the gold standard for stereovision testing. Unlike other tests, the RDS can't be defeated by tricks, such as snaking your head back and forth or rapidly changing where you focus your eyes.

Julesz's invention also influenced a generation of neuroscientists by showing that precise, computer-generated stimuli can reveal the inner workings of human vision. Or, as Julesz put it, "I was able to do psychoanatomy without a knife."[2]

Julesz's breakthrough inspired future Nobel laureates David H. Hubel and Torsten N. Wiesel to investigate stereovision the old-fashioned way—and if you, like me, are a squeamish animal lover, you may want to skip the next few paragraphs.

H&W, as they were often called, were an odd couple. Hubel was a garrulous Canadian with a head of wild curly hair. Wiesel was quiet and serious, a classic Swede. Hubel was beloved by students, who appreciated his ability to explain complicated concepts

---

* Sadly, Julesz could not be reached for comment, as he died in 2003.

in plain English. Wiesel was also beloved by students, but only the very best ones. His garbage can overflowed with work by grad students that didn't meet his sky-high standards of scientific rigor.

The two (who never hung out after work) formed one of the most productive teams in neuroscience history. Throughout the '50s and '60s, they made many major discoveries, including the retinotopic organization of the primary visual cortex and the existence of complex neurons that encode things like lines and movement. They also discovered that Julesz's prediction about stereovision was right — it does happen on a pixel-by-pixel basis, quite early in the visual-processing stream. (That said, triangulation may play a role later in the process.)

In a series of landmark experiments, H&W took newborn animals and sewed one of their eyes shut. A few months later, they reopened the eye, and discovered that the animal was effectively blind in that eye. When they took a peek under the hood, the reason behind this blindness was immediately obvious.

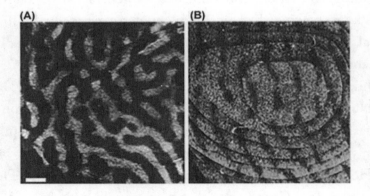

Above are slides from one of H&W's experiments. They show the primary visual cortexes of two different monkeys. The left image is from a normally developing monkey, and the right image is from a monkey whose left eye had been sewn shut soon after birth. The lighter stripes are neurons that linked to the right eye, and the darker stripes are neurons that linked to the left eye. (These stripes are called ocular dominance columns.)

In the normal brain, you can see how the white and black stripes take up about the same amount of real estate. But in the animal whose left eye was sewn shut at birth, neurons that should have been responding to the left eye have switched their allegiance to the right, rendering the left eye effectively blind. That reorganization persisted even after the closed eye was reopened.

In follow-up studies, H&W simulated strabismus in lab animals by severing eye muscles, and they found similar results. Misaligned eyes also cause neurons to shift their allegiance to the working eye. However, neither of these experiments works on mature animals — if you sew the eye shut on an adult monkey and leave it closed for a few months, when you open it up, the eye will work just fine, and their cortex is similarly unaffected.

These results are where we get the concept of critical periods of development — short windows of neuroplasticity, after which the arrangement of neurons becomes locked down. In the case of stereovision, that window is extremely short. Infants start seeing in 3D at about three and a half months. And yet, many children and some adults have demonstrably gained stereovision quite a bit later. Susan Barry, for instance, learned to see in 3D at the age of forty-eight. This is something that almost everyone believed was impossible.

Barry wrote Hubel a letter asking him what he made of her experience, and she was relieved when he wrote back and said that he didn't doubt her at all.

Specifically, Hubel noted that he and Wiesel never attempted to rehabilitate the animals — which would have involved corrective surgery or vision therapy for monkeys and kittens. As a result, we have no idea what happens in the brains of people who learn to see in 3D as adults.

To find out, scientists would have to take high-resolution pictures of someone's visual cortex before and after vision therapy—and the vast majority of MRI machines can't zoom in closely enough to capture what's happening.

However, I've recently come across a paper by a scientist who has come up with a new way to take extremely high resolution images of living human brains, pictures more detailed than anyone has ever seen before. As luck would have it, he studies stereovision.

I decide to give him a call.

When Steve finally finishes work, he takes me driving. It's a beautiful spring evening. The robins are laughing, the frogs are peeping, and the squirrels are apparently suicidal. After I stop short a few times to avoid making rodent pancakes, Steve feels compelled to speak up.

"You can just keep going. They usually get out of your way," Steve says.

Usually? I try to do better, but Steve, once again, falls silent. I can tell that he's thinking something and not saying it, and it's driving me crazy.

"What?" I finally say.

"You're drifting all over the street," Steve says. "Are you having trouble seeing where you are on the road?"

It's worse than that: I'm having trouble *seeing*. Every time my brain toggles between my left and right eye, the whole world jumps sideways—a few inches for close-up things, and a few feet for things that are farther away. Normally I don't notice when I switch eyes, but something about driving is making me conscious of a glitch that my brain usually smooths over.

To stabilize the scenery, I close one eye—and suddenly I realize why lazy eye used to be called "squint."

I'm keeping these problems to myself. Steve is already a nervous wreck; full disclosure would only make it worse. I do, however, have a practical question.

"How are you supposed to center yourself in the lane when you're sitting way over on one side of the car?" I ask.

"You get a feel for where your car is on the road," Steve says. "It's like an extension of your body."

Knowing where your body is in space is actually a big problem for people who grow up stereoblind.[3] Neurotypical children learn how to move by observing their own bodies and correlating that information with messages from proprioceptive sensors in their muscles. Stereoblind children are missing crucial visual information, and the precision of their muscle calibration suffers as a result. Good stereovision also helps when learning to balance, which is why stereoblind people have trouble recovering from missteps and tend to fall. Combine that with the fact that we simply have trouble estimating where other objects are, and it's no mystery why kids who are labeled clumsy often turn out to be stereoblind.

While many doctors will tell you that stereovision is only necessary for fighter pilots and eye surgeons, scientists have found that stereoblindness causes a wide range of deficits in basic visual processes, such as detecting contours, identifying shapes, tracking moving objects, and counting features in a complex scene.

Stereoblindness also causes many practical problems: People with it are slower and less accurate at aiming their eyes—yet another reason why we don't make very good bird-watchers. We have great difficulty identifying objects in clutter, and we can't read words where the letters are squished close together.[4] (This is, of course, hard for everyone, but it's extra difficult for us.)

People who grow up stereoblind develop a variety of compensations, including some that are invisible to the naked eye. Scientists with video cameras have captured, in slow motion, major differences in the movements of stereoblind and neurotypical folks. For instance, when reaching out to grab something, we stereoblind folks move our hands more slowly and make more mid-reach course corrections. Then, when we make contact with whatever we're trying to grab, we really let our fingers linger before picking it up—perhaps using our sense of touch to make up for our lack of depth perception. Scientists have also found that stereoblind people walk more slowly and take shorter steps.[5]

It's not all downsides, though. There are at least two small advantages to being stereoblind. As Margaret Livingstone found, stereoblind artists have a slight edge when drawing or painting. The other advantage only applies to people who have exotropia (outward-pointing eyes) like me. Some of us enjoy a slightly expanded field of view. This is because our brains only suppress the *central* vision of our weaker eyes, while leaving the peripheral vision "on." It's not as good as having eyes in the back of your head, but a handful of exotropic parents have told me that their bonus peripheral vision helps them keep an eye on their kids.

The bottom line? Stereoblind people see the world very differently from neurotypical folks. And I'm afraid that, while learning to drive, I'm going to make mistakes that seem boneheaded to Steve.

As the sun sets on my first driving lesson, I'm finding it increasingly difficult to see the road. This is a big issue, because here in the mountains, roads are often edged with cliffs instead of curbs. After a few close calls, Steve starts giving me regular updates about my distance from certain death—I mean, the side of the road.

"Two feet, one point five, one... Watch out for the ditch! Okay, now you are in the wrong lane..."

This is not going well. Maybe I should have tried to learn to drive *before* buying a house so far from civilization.

9

SADIE VISION

Not to brag, but I am about to slide into a 7-Tesla MRI machine, one of the most powerful in the world. First, I have to get past a metal detector—one that keeps beeping at me, even though I am buck naked beneath my hospital gown.

"Are you sure that you're not wearing any hairpins?" asks Shahin Nasr, a Harvard Medical School radiology professor. He's got a full head of dark hair and movie-star good looks, like an Iranian Brad Pitt. "No underwire bra?"

I blush and I pat myself down. I'm pretty sure I don't own any bras or hairpins, but you can't be too careful when you're in the presence of a magnet like this—more than 140,000 times stronger than the Earth's magnetic field. In the annals of MRI safety history, two incidents loom large: one woman's hairpin turned into a projectile and lodged itself in her soft palate, and a mislaid oxygen tank crushed a child. Those accidents involved 3T MRIs. No one wants to see what a 7 T could do.

Nasr asks, for the third or fourth time, if I have any metallic implants or shrapnel in my body.

"No," I say—though at this point, I'm just hoping it's true.

Nasr works at the Athinoula A. Martinos Center—"the Google of biomedical imaging," as he describes it—and he's using their 7T MRI to capture the human visual cortex at an unprecedented level of detail. A typical 3T MRI machine zooms down to about 3 cubic millimeters—a few million neurons. The 7T's voxels (3D pixels) are 0.5 to 1 cubic millimeter, so they zoom down to 6,000 neurons or so.

What does this mean? Well, normal MRIs can capture brain activity at the level of small structures, such as the FFA. Nasr's 7T can image columns of neurons, such as the ocular dominance columns that H&W observed by making slides of the brains of (dead) animals.

It seems the metal detector is on the fritz, so Nasr waves me through. I stop racking my brain for forgotten childhood shrapnel incidents and concentrate on getting as comfy as possible. Another quirk of the 7T is that it's extremely sensitive to even the smallest movement. Ideally, I'd hold my breath and pause my heartbeat for the next two hours. But since that's too much to ask, Nasr instructs me to lie very, very, very still.

When the machine sucks me in, I feel like I'm rolling up the wall. The magnetic field is messing with my vestibular system. Some people report disorientation, nausea, and tasting metal, but I just feel like I'm in an underpowered rock tumbler. So far, so good.

Nasr's muffled voice is piped into the machine. It sounds like he's asking a question, so I tap the button on the control pad in my hand that means "Yes." I assume I'm saying, "I'm okay and ready to go," and not, "Sure, go ahead and make slides of my brain."

SADIE VISION

The machine thumps like tennis shoes in a dryer. My job is to stay awake and passively watch a video—and to prove I haven't fallen asleep, I have to click a button every time a plus sign in the middle of the screen changes color.* It's a good thing I have something to do, because Nasr's video is even more boring than DeGutis's. Every few minutes, there's a new pattern—lines of various widths and colors gliding across the screen, grids of dots, both moving and still, and probably other things that I can't remember.

After two hours of this, the machine spits me out. "You can move now," Nasr says.

"Oh, thank God," I say, while vigorously scratching my cheek. A minor itch that began an hour ago has since grown into an all-consuming obsession. I only made it through thanks to advice I've heard, secondhand, from alcoholics in recovery: *Don't think about the future. Take it moment by moment. Just breathe.*

"That was great," Nasr says. "You didn't move at all!"

This props up my flagging self-esteem. There are many things I suck at, but I was *born* to play dead.

Adults who have conquered their stereoblindness report that it's a life-changing experience, like seeing through the eyes of a child. The play of light on an ordinary coat takes on the luscious depth of a Dutch-master painting. Trees transform into exquisite three-dimensional sculptures, forests become more expansive, and negative space becomes almost visible. You might find yourself acting silly, whirling around in snowfalls, or stacking stones in streams, just to watch the water ripple through.

---

* Finger twitches are grudgingly allowed.

Learning to see differently can also affect how you *think*. For instance, Susan Barry remembers being impressed when her children looked at an antique clock and immediately understood how all the parts worked together. She was never able to do that, to imagine multiple processes happening simultaneously, until after she learned to see in 3D. It was almost as if her mental capacity had inflated right alongside her visual world.

Like Barry before her vision changed, I often feel hampered by my mind's insistence on staying on straight-and-narrow, if-then pathways. I wish I found it easier to make the lateral leaps of logic that underpin things like jokes and puns. If I could learn to see in 3D, I wonder, could it help me be funnier, or more creative?

This connection between seeing and thinking feels surprising, but perhaps it shouldn't. After all, language is rife with clues that seeing isn't just a matter of photons and retinas. "I see." "He's short-sighted." "She's got a vision for this company."

Many metaphors equate seeing with understanding—which dovetails with research by Doris Tsao and Rodrigo Quian Quiroga, whom we met in chapter 5. They are finding evidence that the brain builds abstract concepts (e.g., a perfect circle) on a foundation of visual experiences (seeing near-perfect circles in nature).* This makes sense, because as visual creatures, seeing is the primary way most of us come to know the world.

It's hard for me to imagine what seeing in 3D would feel like given my current state of mind: all-encompassing ennui. Even before the pandemic, even before I was laid off, I was getting a little bored with my (ridiculously amazing) life. I'd seen all the plays, eaten at all the good restaurants, been to all the exhibits, gone to all the parties. That was my job: to do fun things and then

---

* And probably other senses as well.

write about them. Now that I'm living in the middle of nowhere and unable to drive, I'm learning what boredom really means.

If you're thinking, *Cry me a river,* I agree. But middle-aged blahs are not just reserved for the comfortable. Many studies, including one of two million people from eighty nations, have found a U-shaped happiness curve, with middle-aged people at the bottom, and the young and old gleefully holding up the ends.[1]

Other great apes also experience this U-shaped trajectory. According to a study that involved surveying zookeepers, chimpanzees and orangutans are least happy at the ages of twenty-eight and thirty-five years old—halfway through their life spans.[2] Similarly, humans are least happy when we are forty-five to fifty years old.

All the obvious potential culprits—social and economic status, life demands, having children or not having them—have been ruled out, says Andrew Oswald from the University of Warwick in England, coauthor of both of the aforementioned papers.*

"The U shape is one of the most pervasive and puzzling patterns in modern behavioral science. It also matters hugely for us in our everyday lives," he says. "I hope to find out its scientific cause before I drop off the end of the graph."

Here's my theory: By the time you're middle-aged, your brain knows exactly what to expect, every hour of every day. You know what information to pay attention to and what to ignore. You know exactly how that chocolate cake is going to taste, and so you

---

* I asked Oswald if the U-shaped happiness curve applies to apes in the wild, or to the few existing hunter-gatherer tribes. He said it's an interesting question that no one has tried to answer, as far as he knows. So, if you're a grad student in search of a very difficult topic for your dissertation, one with lots of opportunities for travel, feel free to look into this.

devote fewer resources to actually tasting it. A boring world is a safe world, and our brains evolved to keep us safe.

As for why we get happier after middle age, maybe looming death makes us appreciate ordinary pleasures? I'll let you know, but in the meantime, I'm hoping to recapture my sense of wonder by learning to see the world in 3D.

There are dangers, however. In mature brains, haywire brain chemistry resulting in excess neuroplasticity has been linked with a host of ailments, including neuropathic pain, tinnitus, vertigo, and muscle spasms. If you're ever feeling cranky, hyperactive, or hypersensitive, it may be a sign that your brain is undergoing an unusual amount of change—and this is exactly what happened when Barry began seeing in 3D.

What really gives me pause is Adriana Valencia's story. Like me, Valencia had several childhood surgeries that made her eyes appear aligned, but she remained completely stereoblind. In her twenties, she tried vision therapy—and it worked. After about eight months, the world popped into vivid, terrifying 3D.

"It was kind of scary and unpleasant. I'd be walking and I'd be like, *Oh my God, why are these trees sticking out at me? Why is this grass so pointy?*" Valencia recalls.

Valencia quickly discovered another major downside of 3D life. An Eastern Studies PhD student, Valencia relied on her ability to quickly grok heaps of literature, in both English and Arabic. This ability took a nosedive when she started seeing in 3D.

"I'd read the same sentence, over and over again, and have no idea what it meant," she recalls. "My conclusion is that stereovision was trying to take up a part of my brain where my reading-comprehension skills live."

When Valencia quit vision therapy, her world deflated and her reading comprehension returned. Still, her story gives me pause. Reading is a significant part of my job too. I'm going to feel awfully dumb if I'm the first person to *uncross* my eyes and get them stuck that way.

If I succeed at learning to see in 3D, I also run the risk of giving up that small artistic advantage, but that doesn't bother me. I've had forty years to learn to paint, and there's no reason to think I'm going to suddenly get interested in it.

This brings me to my second theory of middle age. Over the last four decades, I have fully capitalized on my strengths. I play violin in a bluegrass band. I've made a career out of writing and talking to strangers. Recently I've even found a use for my ability to lie still in small spaces. If I want to have new experiences and keep growing, I need to start addressing my weaknesses — including, or perhaps especially, the ones that seem all but baked into my brain.

My first target, improving my face perception, wasn't a resounding success. But learning to see in 3D seems like a more tractable goal — it is, after all, something that has been done before. What's more, improving my stereovision might also improve my face perception. If you make a neurotypical person cover one eye while learning and matching faces, their performance goes downhill fast.[3]

My best chance of gaining stereovision is to go see a vision therapist. Unfortunately, the nearest behavioral optometry center is almost an hour from my house, and my insurance won't cover the considerable expense — two hundred dollars for each weekly forty-five-minute session. To find out what other options I might have, I dial Dennis Levi, a professor of neuroscience and optometry at University of California, Berkeley. When I searched for "amblyopia" on APA PsycInfo, he was the top author, with more than eighty studies to his name.

I explain that I'm looking into new treatments for stereoblindness—to try out and to potentially write about. In a gentle South African accent, Levi delivers some bad news.

"On average, the treatment for amblyopia hasn't changed since 1743 when a French guy, Georges[-Louis Leclerc], Comte de Buffon, figured out that one should penalize the strong eye so that the weak eye can learn to work. And that's still the standard treatment," he says.

That may be about to change. Several companies are developing programs for virtual-reality headsets that teach stereoblind people how to use their two eyes together. Even if your eyes are pointed in completely different directions, VR technology can center the image on each of your retinas and gradually train you to bring your eyes together. VR-based programs can also "juice up the signal" to your weaker eye—making it so bright, the brain has no choice but to pay attention.

Preliminary research on this technology is promising, Levi says, but it's mostly been tested on children. "There's this notion that if you're over seven or eight, your brain is no longer plastic. It was a rationale for not doing or even trying treatment on older patients. And so we've been very interested in seeing if you can teach old dogs new tricks."

"I am an old dog!" I say. "I'd like to learn new tricks!"

Levi tells me to send my eye-exam records to his lab assistant to see if I qualify to join one of his studies. I forward them before we even get off the phone.

I'm staying at Pam's place again. She and her partner (Paul?) recently moved, and they aren't even unpacked, so it's extra nice of them to let me crash on their couch.

Before I ring the doorbell, I consult my "people to remember" email folder, and read the note I wrote to myself more than a year ago. A man opens the door.

"Hi, *David*," I say brightly.

"Hey, Sadie," Pam calls from the kitchen. She's got hot-pink hair now, but I feel confident that she is who I think she is. The role of David, on the other hand, could be played by any tallish white guy with glasses.

I tell Pam that I'm meeting up with the Anns for dinner and invite her to come along.

"I would love to, but I have plans with my sister," Pam says.

"She lives near here, right? That must be nice," I say, proudly displaying another tidbit of information gleaned from my notes. I also know that she's seven years older than Pam, and I'm actively looking for a reason to mention it.

Last time I was going to meet up with the Anns, I bailed and flew home early, so I'm a little anxious about seeing them now. It's been *years*, almost a decade. Will I recognize them? Will they recognize me?

When I get to the restaurant, Brown Anne is already there. Seeing her brings back a flood of memories that I thought I'd lost forever—Tae Bo in the living room, bribing our friend Nora to make out with her hockey stick, pretending that Hartford was Miami when we couldn't afford to go anywhere for spring break. Red Ann arrives shortly thereafter, and she's trailing memories as well: making fancy cocktails in her room, having afternoon tea, almost being crushed by a semitruck. I wonder if my inability to store my friends' faces in my brain means that I need to see their faces to access happy memories. (I make a mental note to buy a digital picture frame.)

I ask the Anns if they have any good gossip, and they do. Sadly, I have no idea who they are talking about. Instead of bluffing, I confess.

"I hardly remember anyone from school anymore," I say. "Honestly, I don't even remember your kids' names, or how old they are, or where exactly you live."

"You've always been like that," Brown Anne says.

"No one expects you to remember things," Red Ann says. "We expect you to tell good stories."

My heart floods with joy and gratitude. These guys really know me! And while the facts of their lives slip through my fingers, I know them too—amorphous things, like the cadence of their speech, the kinds of drinks they will order, and how they will respond to my stories about learning to drive. (Red Ann: concern for my safety; Brown Anne: concern for other people's safety.)

When I get back to Pam's, I realize that I failed to take any notes, and I've already forgotten half of what the Anns told me. But I think that's okay. When you're with your very best friends, you can be your most authentic, unvarnished self.

Self-acceptance—what a novel idea.

Before my third day of scans, I meet up with Nasr in a waterfront park dotted with old industrial buildings. I explain my overall project and my hunch that my faceblindness was caused, in part, by my stereoblindness.

"What do you think caused your amblyopia?" he asks.

"Well, since that doesn't run in my family either, I'm thinking that I probably experienced some birth trauma, maybe hypoxia," I say. "My mom says that she had a super long labor and I came out a little blue. And then I had those eye surgeries that didn't really fix anything..."

"So you blame your parents..." Nasr starts to say.

This logical leap leaves me briefly speechless.

"I don't blame anyone!" I say. "I just think that maybe my head was too big, or whatever..."

Nasr says that if he were my dad, he would be very stressed out by my project. "I can't imagine if my son were writing a book about me, saying, 'You know, my dad was a cruel person...'"

"No, no," I protest. "My dad was great—both my parents are great. They had very little money and still managed to get me the best health care available..."

Nasr has two young kids, and it seems I've got him worried that they are going to call him in thirty years with questions.

I change the subject to Nasr's research. I already know from internet sleuthing that he's mapping the human visual system at an unprecedented level of detail—finding substructures that we suspected existed in humans but we only knew from research on monkeys.

Remember the primary visual cortex? It's also known as V1, and the next steps in visual processing are creatively named V2 and V3. From monkey research, we know that these areas contain thick stripes, which process binocular disparity (3D vision), and thin stripes, which process color. Nasr's groundbreaking research has shown that this is also true for humans. Furthermore, in stereoblind people, neurons in the thick (3D-processing) stripes switch jobs and start processing color.

"Oh, cool!" I say. "Do I have super color vision, then?"

"No," says Nasr, who will rain on my parade several more times before we're done chatting, but I won't mind, as he's quite charming.

Nasr explains that amblyopia is actually associated with color vision *deficiency*, and this fits with my personal experience. While I am not colorblind, I have long suspected that I see color differently than other people do. This once resulted in a newsroom-wide vote on whether or not a vest I described as orange was actually yellow. I lost in a landslide, but I couldn't shake the feeling that all my colleagues were playing a trick on me. (I mean, the vest was obviously orange.)

Another amblyopia deficiency that Nasr is investigating is our difficulty discerning details at a distance. This isn't simply an acuity issue: I have 20/20 vision (thanks to LASIK), but I can't read signs that other people read with ease. Why? One possibility is that people with normal vision learned as babies to devote more neurons to analyzing things that are farther away. Since stereoblind babies can't judge distance very well, we miss that lesson.

"So, even if I learn to see in 3D, I'll probably still have all these downstream deficits?" I ask.

Theoretically, yes, says Nasr, and he follows that with more bad news: amblyopia probably can't be cured in adults for the simple reason that it would require a fundamental change in the basic architecture of our visual system. The foundational aspects of perception are locked in by molecular gates — and while neuroscientists are trying to figure out how to reopen those gates to learning, we are decades away from trying this with humans.

I tell Nasr about Susan Barry and other cases of adults who learned to see in stereo, but he isn't swayed. He, and many other scientists, suspect that these people were never truly stereoblind — that as babies they had some glimpse of the 3D world they were later able to build on. My feeling is that even if this was the case, it would mean that a lot of apparently stereoblind adults could potentially learn to see in 3D, maybe even me.

Even though he's pretty sure it won't happen, Nasr agrees to rescan my visual cortex if I learn to see in 3D, so we can see what, if anything, changed at the neural level.

SADIE VISION

"Why do you want to see in 3D?" Nasr adds. "Your vision is perfect. You have Sadie Vision! No one else has vision like you."

This "Buck up, kiddo" message reminds me of my dad's point: don't worry about something you can't change. It's a kind impulse, but it strikes me as disingenuous. I mean, if being stereoblind were so great, the NIH would not be spending tens of millions of dollars on amblyopia research annually.

Also like my dad, Nasr is anxious about me driving. "So, why do you insist on driving?" he says with a nervous laugh. "Because this is one of those things—it's maybe dangerous."

"It's actually safe for stereoblind people to drive. I'll send you the studies," I say.

I never send him anything. Parental anxiety is like parental guilt—it just can't be reasoned with.

10

HOLLYWOOD MEETS SCIENCE

In December 2009, James Blaha did not want to see *Avatar* with his family. "3D movies never work for me," he said. "They just give me headaches." The freelance computer programmer had been stereoblind his entire life—a condition he felt was partially his own fault, because he was not very good about wearing his eye patch as a kid.

His parents talked him into going anyway. A few minutes into the film, one of the two projectors stopped working. For everyone in the theater with normal vision, the movie turned into a regular 2D film. From Blaha's perspective, however, the screen turned black. Then he noticed a sliver of the movie was still visible at the far left of the screen. When he closed his right eye to force his left to work, the entire screen filled back in.

"The way the projector was glitching showed me a map of my suppression," Blaha says. "It was much more interesting than the movie."

Blaha had no idea what was going on with the blue aliens, but he was gaining new insight into how different his vision was from most people's. His brain was ignoring all the information from his nondominant eye—except for that little slice in the periphery.

As he left the theater, Blaha began formulating an idea to create a 3D video game that would teach his brain to integrate information from both of his eyes into a single stereoscopic image.

The moment Blaha got home, he looked into buying a 3D projector and was dismayed to find they ran several thousand dollars. A few years later, virtual-reality headsets began to hit the market, and Blaha bought himself an Oculus Rift developer kit for a few hundred dollars.

For his first experiment, Blaha created a 3D rotating cube that allowed him to adjust how bright the image appeared to each eye. To overcome his brain's suppression, he made the image presented to his left eye brighter and brighter. Once it was about twenty times brighter than the image presented to his right eye, something unexpected happened. The cube popped out at Blaha, and he jerked his head backward to dodge its pointy corner.

"I was really surprised," he recalls. "I'd never seen anything like it."

Blaha's next experiment, Breaker, was inspired by the classic game where players hit virtual ping-pong balls at grids of squares.

He also began working on a virtual prism, which could center the image on each of a person's eyes, regardless of their alignment. The idea was to gradually reduce the prism in order to train people to use their own eye muscles to keep their eyes aligned. This,

Blaha says, would be much better than surgery, since it would improve people's vision, rather than just be cosmetic.

This was all sucking up a lot of Blaha's time, so to fund the project, he created an Indiegogo campaign. It was an instant hit. Within two days, Blaha raised more than twice his initial goal of two thousand dollars. Two months in, he had twenty thousand dollars.

"That was when I realized how many people have lazy eye, and what a big demand there is for this kind of technology," Blaha says.

With the help of a startup accelerator, Blaha created a company called Vivid Vision and hired a few computer programmers and a scientific advisor. The original plan, he says, was to sell Vivid Vision directly to consumers. But after a few beta testers experienced moments of double vision, Blaha (or perhaps his lawyers) decided the program was too powerful to be used without medical supervision.

The FDA approved Vivid Vision as a medical device in 2018, and vision therapists began offering the program through their offices. However, since the regulatory agency is still developing its policies around "digital therapeutics," it's been a bumpy road. On January 1, 2022, the company had to temporarily stop selling the software in the United States to work through some new FDA requirements.

Blaha is using this as an opportunity to add automation, so that vision therapists won't have to constantly adjust the settings for their patients. "We're writing algorithms that automatically determine what level each patient is at, for each visual skill, and then it auto-configures all the activities," he says.

Blaha isn't just the founder of Vivid Vision; he's also a client. After about a year of playing his own video games, Blaha went hiking and noticed a tree branch was "sticking out" at him, in an accusatory manner.

"I was shocked," he says. "It's hard to describe in words what it's like."

Once he got over his surprise, Blaha marveled at how different the forest looked in 3D. Impenetrable tangles of branches became airy and open. For the first time, he could really *see* the spaces between trees.

"I don't know about movies, but the real world looks better in 3D," he says. "It's crisper, more vivid."

Blaha is eager to get his invention into the hands of stereoblind people, but it's been a real slog. On top of the FDA's dithering, there's the constant chore of proving Vivid Vision's potential profitability to investors. "They're playing this game of optimizing revenue and optimizing products," Blaha says, "and it's really easy for them to forget that this is not a game — it's actually about real people doing real things for their health."

Another VR-based therapy company, Luminopia, pulled ahead of Vivid Vision on October 20, 2021, when they got FDA approval to use their VR software to treat amblyopia in children ages four to seven. In a randomized, controlled clinical trial, Luminopia's software improved about two-thirds of the kids' vision, so that their weaker eyes were able to read two additional lines on the classic eye chart. (Interestingly, the control group improved as well, by about 0.8 lines.) The experimental group also showed a slight, statistically significant improvement in their stereovision — and these were kids who had already been failed by traditional treatments, such as patching the strong eye to make the weak eye work.[1]

Unlike Vivid Vision, Luminopia doesn't use video games — instead, kids simply watch television shows while wearing VR goggles. The Luminopia software degrades the image presented to the stronger eye while showing the weaker eye a clear, high-contrast image. This is so annoying, the brain starts using the weak eye to complete the image.

"This is the first new development in amblyopia treatment in the last decade," says Endri Angjeli, Luminopia's vice president of clinical development.

I beg Blaha and Angjeli to let me try out their software. I emphasize the free publicity they'd get. I tell them about my impulsive move to the middle of nowhere, how desperate I am to learn to drive, and how seeing in 3D might give me a leg up. Both Blaha and Angjeli are sympathetic, but they can't help. If they want to stay in the FDA's good graces, their hands are tied.

After I get back from Boston, driving lessons resume, and they are...stressful. Every time I take the wheel, my usually stoic husband radiates so much anxiety, it's nearly visible. Meanwhile, my face, according to Steve, is "a rictus of fear."

I'm as scared as I look. You know how parallel lines converge at the horizon? For me, this results in an optical illusion whereby oncoming vehicles often seem to be heading *right at me*. Every time this happens, I scream—and it takes every ounce of willpower not to swerve off the road.

When I ask Steve how he manages this challenge, he turns a new shade of gray. I guess this isn't a problem for him.

After a month on empty country roads, Steve decides it's time for me to try a two-lane highway. I hit the gas, and the truck accelerates to a breakneck fifty-five miles per hour. It's clattering along so loudly, I fear we're shedding parts. Also, why do I keep getting passed?

"Why is this happening?" I shout. "I'm going the speed limit."

Steve explains that the *real* speed limit is about five miles per hour faster than the posted one.

This is truly shocking. "Are you telling me that everyone is *breaking the law*, all the time?"

I can't turn to look, but I suspect that Steve is rolling his eyes.

When we arrive at the grocery store, it takes me three tries to park to Steve's satisfaction. Deeply relieved, we stagger into the store and buy ourselves celebratory sodas.

"Are you ready to drive back?" Steve asks.

"Can you do it?" I plead. "I've had enough excitement for today."

Steve takes the keys, but another problem soon presents itself: the truck isn't starting. Not to get too technical, but whenever he turns the key, the engine goes *click, click, click,* instead of *vrooom.*

"The battery's dead," Steve says. "We need a jump."

Before we even ask for help, a guy in a camo trucker hat appears with jumper cables in hand. Steve tells me to come observe the procedure, and Camo Hat agrees.

"With a truck this old," he says, "you're going to be doing this a lot."

He was right. When I wake up the next morning, the battery is dead, and I have to jump it and drive around for an hour to charge it up. The morning after that, it's really dead. No amount of jumping can bring it back to life.

Then the starter expires, followed shortly by its friend, the alternator. The bad news: we are spending a lot of money at the mechanic's. The good news: I'm becoming increasingly adept at driving there.

Then things settle down. A month passes without incident. "I knew it would work out," I tell Steve. "We just had a rocky start."

Little did we know that the truck was merely biding its time, lulling us into a false sense of security.

One afternoon, Steve goes on a Home Depot errand, and comes home sweaty and wild-eyed.

"The accelerator got stuck," he says. "I was on the highway and it just went faster and faster and faster..."

"Oh my God, what did you do?" I reply. "Did you put on the parking brake?" (Somewhere I had read that this was the thing to do.)

"No, I stomped on the accelerator, and it popped back up," he says. "But I think we need to get rid of this thing. It's dangerous!"

I'm reluctant to let go of my dream of hauling dirt and kayaks in the bed of an ancient truck. Sure, she almost killed my husband, but the same could be said for me. I know Steve's right, though, so I let him take me to a normal used-car lot—the kind that has dancing tube men and plenty of gas on hand.

We loiter by a Mazda3, and a salesman comes right over, copies my license, and hands me the keys.

"Well, that was efficient," I say to Steve, a little disappointed.

I'm deeply anxious about potentially damaging such a pristine, undented car, so I proceed very slowly and carefully. As I turn onto a small road, a warning flashes on the heads-up display. "Turn wheel!" it says. Leaving nothing to chance, an animated steering wheel helpfully demonstrates which direction I should turn it. "Come *on*," I whine. "I wasn't even that close to the curb."

I flick on my blinker, which causes Ms. Mazda to bleat and flash lights, alerting me to the fact that there's someone in my blind spot.

"I see him," I say. "Chill out."

This car is bossing me around, and I love it. I feel like a

petulant teenager, which is a big improvement over being a terrified old lady. A hot little car with the soul of an anxious parent — this is the perfect vehicle for me.

We buy the Mazda, and it's not long before I'm driving everywhere — the park, the grocery store, the other side of the mountain, just to see what I can see. This car is so much easier to pilot than my truck. The driver-assist features are great, but even better is the fact that, when I turn the steering wheel, the car actually changes direction. This was not always the case with the Ford.

It's not all smooth sailing — staying centered in the lane continues to be a challenge. I conduct some internet research on the topic, and I keep finding the same advice: align your hood ornament with the right lane line. Ms. Mazda doesn't have a hood ornament, so I buy her one — actually, it's just a big bulging refrigerator magnet, but it does the trick.

In another internet-inspired innovation, I order magnetic bumper stickers for the front and back of my car. They read, STUDENT DRIVER: PLEASE BE PATIENT.

"Got a teenager, do ya?" a man asks at a gas station.

"Actually, *I'm* the student driver," I say. "I never learned when I was a kid."

"Well, best of luck to ya," he replies with a chuckle. As I navigate out of the station, I notice that he's giving me a wide berth.

I'm usually by myself on these adventures, which is technically legal, since I do have a valid driver's license. I've been renewing my questionably obtained license since 1996, successfully transferring it to DC and to West Virginia without anyone so much as batting an eye. If you find this alarming, please remember that I live in the middle of nowhere. I am not endangering anyone but myself. (And the occasional squirrel.)

My driving record continues to be, technically, perfect. My

parking record, however, is a little spotty. Steve backs his car into the garage without so much as slowing down. I, on the other hand, take extreme care—meticulously lining myself up and then backing into the garage at about 1 mile per hour. I stop every few feet to check my alignment—and I still scrape the sides of my car on a regular basis.

The first time this happens, I almost knock my mirror clean off. That incident leaves a few little marks on the car's right flank, but they are basically invisible due to copious amounts of dirt and dust. No need to report this little incident to the husband, I decide.

My first (and only!) true collision happens when I hit my old truck in the driveway. Not to blame the victim, but Ms. Mazda didn't make a peep, so I assumed we had enough space to get by. We didn't. I tear the truck's bumper half off, seriously depleting its already minuscule resale value.

Since it's unlikely that Steve will fail to notice the truck's bumper dragging in the dirt, I confess. This spurs him to give me a useful bit of advice, something I wish he'd told me months ago: "You don't have to back into the garage, you know. You can drive in front first."

I haven't had an accident since that one. At least, not as far as you know.

After my second surgery, my eyes seemed aligned most of the time, but not always. When I got tired, my left eye would drift off to the side. It didn't occur to me then, but now I am wondering if my wonky eyes are also part of the story of why I felt left out by other kids.

More urgently, my eyes have drifted apart in recent years, and reading about strabismus has me worried about what kinds of assumptions people make about me today, due to my misaligned eyes. A quick dip into the literature does little to assuage my fears.

Teachers rate kids with misaligned eyes as less intelligent, lazier, and less happy than their classmates.[2] Children are less likely to invite kids with misaligned eyes to their birthday parties.[3] Give a kid a cross-eyed doll, and they will be less likely to kiss it and more likely to hit it.[4] Mothers of walleyed kids are less nurturing and more dictatorial than moms with straight-eyed kids.[5] These moms also tend to be more depressed. (No word on the dads in that study.)

The stigma of misaligned eyes doesn't let up when you reach adulthood. Studies show that people with misaligned eyes are less likely to be hired or considered for promotion.[6] We are also viewed by potential dates as unattractive, unintelligent, and bad at communication.[7] (The "bad at communication" assumption may be rooted in the importance of gaze direction and eye contact as social cues.)

All this bias really weighs on people: strabismic adults say they would trade an average of five years of their life span for cosmetically aligned eyes.[8]

Reading about strabismus stigma causes me to start seeing it everywhere. Remember Ed, the crazy hyena henchman in *The Lion King*? All he does is laugh like a loon, and his misaligned eyes further signify his derangement. *The Simpsons* character Ralph Wiggum, famous for bon mots like "My cat's breath smells like cat food," has subtle exotropia, which enhances his clueless look. In one episode of *Rick and Morty*, a copy of Rick comes out goofy and walleyed. Everyone's favorite walking id, *Sesame Street*'s Cookie Monster, often has googly eyes. (To be fair, old-school Cookie Monster accurately captures how I act around dessert, but I can see how other people might feel misrepresented.)

HOLLYWOOD MEETS SCIENCE

While exotropia (walleye) tends to telegraph craziness, esotropia (cross-eye) often signals that a character is temporarily overwhelmed—for example, Austin Powers after getting hit in the crotch or while thinking too hard about time travel. A variation on this theme is a cross-eyed "O" face called ahegao, which originated in Japanese porn, and characteristically includes drooling or other signs of losing musculoskeletal control. On TikTok and Instagram, going cross-eyed has become a handy way for gorgeous influencers to look silly and more approachable.

These backward attitudes are baked right into our language. As you may have noticed, "cockeyed" means "foolish" or "absurd."

Given the pervasiveness of strabismus stigma, it's odd that I, personally, have never noticed anyone looking askance at my misaligned eyes.

When I bring this up with the Anns, they claim that I've never cared about my appearance.

"That's not true," I say. "I'm as vain as anyone!"

As I try to square these two opposing facts, an alarming hypothesis begins to emerge: Unless there's immediate evidence to the contrary, I assume that I look somewhere between acceptable and fantastic. Often I am wrong.

I remember one particular instance when I was in the Metro, and a frowning man pulled me aside and handed me a list of nearby homeless shelters. "Oh," I said. "I'm not homeless."

"It's true," Steve added. "She lives with me."

"Thanks a lot," I whispered. "Now this guy thinks I'm your charity case." (Steve wisely refrained from pointing out that, at the time, this was basically true.)

Wondering why this guy assumed I was indigent, I took a look at my reflection in the Metro car window and noticed a few small, potentially relevant details: I had a smudge of dirt on my face, my hair was a bird's nest, and I was carrying a garbage bag full of my belongings.

Why hadn't Steve told me to wash my face, or brush my hair, or come up with a different solution for keeping my backpack dry in the rain? When I asked him, he said he hadn't noticed me looking particularly disheveled. This leads me to two equally alarming possibilities: either Steve is wildly unobservant, or I always look this way.

Is my overly optimistic self-image a stereoblind thing? Or a faceblind thing? I ask the relevant Facebook groups if anyone has had similar experiences, and all I hear are crickets.

Later that summer, my dad, stepmom, and grandmother drop by for a visit. While Steve tends to dinner, my dad (whom my friend Sybil describes as "a non-creepy Bill Clinton") putters around my house, cleaning ducts, oiling hinges, tightening cabinet doors. Like me, he's good at napping, but he's surprisingly bad at sitting still.

Steve and I, on the other hand, have been loafing quite a lot. Our post-move unpacking really hit a wall when we discovered that unopened boxes make perfectly serviceable tables and chairs.

"Dad, can you help me build some furniture?" I say.

"Sure," my dad says, even though he knows that I'm really asking *him* to build the furniture. This isn't just a case of me being lazy. I can't hit a nail on the head—and now I know why.

Even with my questionable "help," my dad makes quick work of an entire outdoor furniture set: a table and six chairs. It's too buggy to eat on the porch, unfortunately, so we sit down at the indoor table (which, thankfully, came preassembled).

Steve serves up the brisket, and my family falls uncharacteristically silent. Then come the naked moans of pleasure.

"Oh, Steve," my grandma gasps. "Where did you learn to cook? You should start a restaurant!"

Steve's blushing, so I helpfully draw the attention back to myself.

"I just got into a stereoblindness study at Berkeley," I say. "I'm going to do vision therapy using virtual reality. And if they can cure me, well, watch out, Roger Federer."

"Do you think you were bad at tennis because of your lazy eye?" my dad asks.

"Yes, definitely! It's hard to hit the ball when you have no idea where it is."

I've told my dad this before, but for the first time, he seems to be actually taking it in.

"You never connected my eye problems to my poor showing in tennis and softball?" I ask.

"I don't know. Maybe it's genetic," my dad says.

When my dad starts making self-deprecating jokes, I know he wants to change the subject. But I press on.

"That's not true. You're super athletic. You played rugby and racquetball in college."

"Maybe it's your mom," my dad says.

"No, she's great at tennis," I hit back. "And Saul is a pétanque champion."

"Gosh," my dad says. "Maybe it *was* your eyes."

I can almost see the cartoon light bulb over his head.

Why did it take my dad so long to accept this obvious truth? Maybe all parents want to believe their kids are perfectly healthy and normal. I get it; it's a useful fiction. But I need him to acknowledge reality and accept me for who I am: a person with a brain that is weird enough to interest three different teams of scientists.

"No one in our family has lazy eye besides me. And other than genetics, the most common cause is hypoxia during birth, which causes minor brain damage," I say. "Mom was in labor for a while, right? Do you remember if my heart rate went down?"

My dad's face crumples, and his eyes get shiny.

"I have to tell you something," he says.

It's pin-drop quiet as my dad tells a story about how, when I was about six months old, I fell asleep on his chest while he was in a rocking chair. My dad fell asleep too, and when he woke up, I was at his feet. Apparently, I'd slid down his entire body and landed on the floor.

"You weren't fussing or anything," my dad says.

"Dad!" I say. "That has nothing to do with anything. I was *born* with crossed eyes."*

I walk around the table to give my dad a hug. Nasr was right! Parents are absolutely racked with guilt.

My dad isn't done. He launches into another confession. When I was about a year old, I was scooting around my grandma's kitchen in a rolling baby walker. No one noticed when I toddled out of the room and tumbled down a short flight of outdoor stairs. When my dad and my grandma noticed I was gone, they found me hanging upside down, still in the walker, apparently unbothered by my new perspective.

"We never told your mother," my dad says, a hangdog look on his face.

---

* This is what everyone says, though it's highly unusual. Amblyopia usually shows up after a baby is four months old, or later in childhood.

"Dad—that might have happened *because* of my lack of depth perception, but it didn't cause it."

"He's told me those stories before," my stepmom says. "It's really been weighing on him."

For forty years? My poor dad!

"You're the best dad in the world," I say, squeezing him tight. "I came out great, Saul came out great, we're all doing great!"

I want to add that my recent investigations into my weird brain are an attempt to clear up some long-running mysteries about my life. I'm not looking for culprits, just causes and effects, or maybe an inflection point or two. And I've come to the conclusion that there was one particular moment that changed my life dramatically for the better, and it's entirely thanks to my dad.

"Remember when we ran into Susan at the little grocery store? Whaley's or something?" I say. "And you told me that I could pretend to know people if I didn't know them? And that all you have to do is get people talking about themselves and they will think you are the most interesting person they've ever met? That was the best advice. It changed my life."

"I didn't think you or your brother ever listened to me," my dad says.

"You're not a villain in this story. You're the hero," I say.

Now all of us are in tears—except for Grandma, who has misplaced her hearing aid.

"Can I tell my mom about the stairs, though?" I ask.

My dad makes a scared face, which is hilarious, because they divorced twenty-odd years ago and live hundreds of miles apart. I'm pretty sure she'd just find it funny, but I'll give my dad a break on this one. He's earned it, poor guy.

VIDEO GAME THERAPY

THE CAMPUS OF UC BERKELEY IS A BLAND PASTICHE OF institutional buildings set amidst almost otherworldly natural beauty. I imagine university architects surveying these rugged hills and throwing up their hands — why even try to compete? I shouldn't be so judgmental, though. Berkeley may be fifty shades of beige, but at least it's not Stanford.

I'm standing outside Minor Hall, snarfing down spring rolls, when vision scientist Jian Ding appears at a fire door. I know he's the guy I'm waiting for because he's wearing a blue shirt that says LEVI EYE LAB on it. It's pretty dorky for a lab to have T-shirts. Also, I want one.

We ascend three flights of stairs and pass a poster advertising a study that seems a lot like the one I'm participating in. "Have you or someone you know been diagnosed with lazy eye (amblyopia)? You might be eligible to participate in a vision treatment." It also promises fifteen dollars an hour, which I'm excited about, as it will help fund my new bubble tea addiction.

Three hours later, I realize that fifteen dollars an hour is way too low — these tests are pure torture. I know I also said that about DeGutis's study, but I take it back. The blinding, hallucination-level migraine I develop at Berkeley makes my Boston headache a fond memory.

It all begins innocently enough. Reading lines of letters at a distance, picking letters out of blobs of color, identifying which cartoon character pops out of the blue stereovision card. Everything is fine until Ding asks me to place my chin on a little plastic shelf.

"I should tell you, I sort of freak out on puff tests," I say.

"This is different," Ding replies. And it is. Ding's device is like a stereogram, but there are adjustable mirrors that allow me to make up for my eyes' misalignment. Behind the device is a computer screen that, right now, is showing a horizontal line to my left eye and a vertical one to my right.

"Adjust the mirrors until you can see the plus sign," Ding says.

Someone with normal vision would already see the plus. I, however, can see only one line or the other. Doesn't Ding understand this? It's why I'm here!

Ding continues: Once I see the plus sign, I should press the space bar. The plus sign will be replaced with a field of shapes that look like monochrome beach balls. If the balls on the top of the screen seem closer to my face than the bottom, I press the up arrow, and if it's the opposite, I press the down arrow.

That seems simple enough, except for the fact that no matter what I do with the mirrors, I cannot see the plus sign.

"Just do your best," Ding says.

He turns off the lights and steps into another room. For about ten minutes, I mess around with the mirrors. When that doesn't work, I try closing my right eye. Once my left eye kicks in and I can see the horizontal stripe, I slowly reopen my right eye. I'm trying to keep seeing the horizontal line while the vertical line gradually appears. And I do it! The plus appears, and I am astounded at my newfound ability to *consciously control what I see*. I'm only able to hold it for a second, though, before my left eye taps out.

Making the plus sign appear requires a level of focus and concentration that seems on par with what you'd need to do to levitate small objects. When I look away from the contraption, I half expect to see pencils and paper clips hovering all around me.

Unfortunately, the plus sign is just the setup for the test. The second I hit the space bar, I lose my concentration and I hear imaginary pencils and paper clips clattering to the ground. I'm faced with the beach-ball screen, and I have to choose at random.

After a half hour of this, the screen fills up with text that looks like a computer program. I don't understand most of it, but I can pick out ".49." That's my accuracy—49 percent. I may as well have been flipping a coin.

"Done!" I shout to the next room.

The next test is exactly the same as the first. The only difference is the size and number of the beach balls. My score: 51 percent.

More variations on this theme follow, and at some point I close my eyes while clicking away. My score does not suffer. Random is random.

Later that day, I meet up with Nae, a disability-rights activist who has cerebral vision impairment (CVI). CVI describes any sort of vision impairment that originates in the brain, so it technically includes amblyopia and prosopagnosia—though the term is usually applied to more severe visual deficits. Nae, for instance, is functionally blind. They see patches of light and dark, but that information is almost entirely useless. A dark patch on the ground could be a shadow or a bottomless pit. Nae can only figure out the difference with the help of a cane.

When Nae was a child, they were fully expected to navigate the world like a seeing person—and it was terrifying. Sometimes they got hit in the face with a ball, or walked into a wall. Nae spent all their energy pretending to see: memorizing the layouts of rooms and even the items on vision tests. Nae got particularly good at pretending to fixate on people's eyes. "If I didn't, I'd get yelled at or worse," they say.

Nae's parents knew that their kid had some kind of mysterious brain-based visual impairment, but they refused to accept that it couldn't be fixed. As a result, Nae was subjected to hundreds of hours of vision therapy. Some kids with CVI might not have minded, but it was an ordeal for Nae. "At one point, I was going thirteen hours a day!" Nae says. "I learned some neat tricks, like how to dilate my eyes on command, but it was torture."

"Oh my gosh!" I say. "I just spent three hours being asked to see things that I physically could not see, and it was awful. I feel so bad for you."

I know that my frustrating experience in The Levi Lab is hardly comparable to the nightmare that was Nae's entire childhood—but boy, it was *not* fun. I have two more days of testing to go, and I really want to ghost.

Nae has a message for parents of kids with CVI: Follow your child's lead. Some kids will want to learn to see as normally as possible, but others will prefer to explore the world through different senses. Either option is perfectly fine.

"People know what works best for them," Nae says. "Even children."

"I think that's true for all neurodivergent people," I add. "Some people will want to learn how to function 'normally,' and others will want to focus on their unique strengths."

"And some people will want to do a mixture of that," Nae says. "The important thing is for kids to get to decide for themselves."

Nae says many people with CVI don't have any obvious brain damage, and so they can't get official recognition of their blindness or access to services that could transform their lives.

"That is so dumb," I say. "What's with the gatekeeping? If someone says they can't see, we should probably believe them."

On my last day of testing, I receive my prize: a VR headset! It's not mine to keep, of course, but we will be fast friends for the next two months. My job is to play the Vivid Vision games for about forty minutes a day, six days a week.

First, however, I must demonstrate that I can use the headset without breaking it. A research assistant who introduces herself as Hilary Lu gingerly places a Meta Quest (formerly known as Oculus Rift) on my head, and tightens the Velcro straps. Then she guides me through setting a stationary boundary that is free of obstacles.

"Can I play the games now?" I ask.

Lu presses a button, and suddenly there are four neon-green balls floating before my eyes. I am shocked. Is this what 3D looks like in the real world? The negative space between the balls is... thick! Palpable!

"I'm just going to adjust this real fast," I say. Actually, I'm overwhelmed and I need to return to flatland for a minute. I pull up the headset and look around the room. The air is so insubstantial and thin! I take a breath and pull the headset down, and I'm stunned, once again, by these plump computer-generated balls.

"Click the one that's closest to you," Lu says.

This is not actually one of the games to improve my stereovision — it's a test, one that I will take before and after each training session.

On average, adults have a stereoacuity of 45 arc seconds — and with that level of 3D vision, you can detect fractions of centimeters at close range. You must have a stereoacuity of less than 25 to be an Air Force pilot. Anything over 80 is generally considered impaired.

Want to guess my stereoacuity? Name a number.

Go higher.

Higher still.

Now quadruple that.

You got it: 3,207 arc seconds.

It's not a surprise, exactly, to find out that I'm stereoblind, but quantifying it makes the fact feel more concrete than before. I can't see depth — like, *at all*. No wonder I can't catch a dang Frisbee!

I head home with the headset, which is loaded with a version of Vivid Vision specifically created for The Levi Lab. It includes six games: Barnyard Bounce, Breaker, Pepper Picker, Bubbles, Jump Duction, and Target.

Target is just a basic target-shooting game. In Breaker, you use ping-pong paddles to hit a red ball toward a wall of wooden squares. If you hit a wooden square, it disappears, and the goal is to clear all the squares. For some reason, Breaker takes place in an asteroid field, while Target is at a state fair. In many Vivid Vision games, you win farm animals as you go.

While these two games are derivative, Barnyard Bounce is unlike anything I've ever seen. The goal of the game is to bounce up a series of floating islands—land masses that are either porous or solid, depending on whether you are going up or down.

The task is easy, and on my first try, I bounce myself all the way to level 49. Then something weird happens: the platforms all disappear, and autumn leaves and golden eggs fall from the sky. I can no longer bounce, so I just sit there, watching the scenery, assuming that this little movie is a reward of some kind. Then I fall back to the bottom island.

For weeks, I assume that the golden egg interlude means I've won the game. Then one evening I notice that after the movie, if I'm really fast, I can get over to the platforms and start bouncing up again.

I bounce all the way to level 99, where I am again faced with the golden eggs. This time I realize that I can move my chicken from side to side.

"Steve," I shout. "You won't believe this. You're supposed to *catch* the eggs!"

"Mmmmm," Steve says.

"And now I have a new character to play—a sheep!"

"Neat," Steve says.

"All I had to do was defeat the farmer and feed him to the pigs."

"Mmmmm..."

"You aren't even listening!"

"I am," Steve says. "You had to kill the farmer. It's like an *Animal Farm* game."

"That's not actually true, but I did win a sheep."

Later, when I ask Blaha how he landed on the theme of "Celestial 4H," he just laughs.

* * *

The ladies' hiking group I joined when we first moved to West Virginia has gradually transformed into a stationary gossip club. I enjoy gabbing with my new friends, but it doesn't count as exercise, so I've started hiking by myself.

When I mention this to my dad, he freaks out.

"What if you get lost?" he says. "What if someone murders you?"

I assure my dad that the homicide rate in my local park is quite low. It's also unlikely that I will become disoriented, as I hike the same trail almost every day.

"It's a loop," I say. "You can't get lost on a loop."

A few days later, I prove myself wrong. After bending over to tie my shoes, I get turned around and end up backtracking. This adds a mile to my route, and it convinces me to add another diagnosis to my ever-growing list: topographical agnosia.

The Hovermale trail starts at a bridge and then follows a burbling stream through rocky terrain. After crossing a bog on narrow planks, you traverse a beautiful, moss-covered hill. On the downslope, you hop some rocks across a small stream, cross some more soggy land, and then you're back where you started.

Over the last year, I've found all sorts of cool critters along this route: endangered wood turtles, bright-orange salamanders, warty toads. But I haven't found many animals since my trip to California. I think it's because of my stereovision quest. All I can think is: *Am I seeing in 3D? How about now? Now?*

According to Susan Barry, there's nothing subtle about the moment your vision switches to 3D. For her, it happened after eight months of vision therapy. One day, she got into her car, and her steering wheel popped out at her.

"It was an ordinary steering wheel against an ordinary dashboard, but it took on a whole new dimension that day," Barry writes.

She closed an eye and watched the steering wheel go flat, and then opened it and it popped out again. So that's what I'm doing—whenever I think I see something in 3D, I look at it with one eye closed and then with both eyes open. While I do feel like I glimpse some unexpected volume from time to time, the moment I close an eye, the world goes flat and gets stuck that way.

These moments of stereoscopic vision are so flimsy and ephemeral, it's possible I'm just imagining them.

Further evidence for the null hypothesis is that I am not getting any better at spotting birds in trees. Indeed, I am getting *worse* at bird-watching. I recently went looking for fall warblers with some friends, and it was torture.

"Oh, look," Steve said, finger pointed in the direction of about a billion branches. "It's a pine warbler."

"I see it!" my friend Sara chirped. "Oooh, it's so pretty."

"Do you see it?" Steve asked.

"Not yet..." I said.

"Okay, do you see the branch with the V, the one behind the tree with the berries..."

I knew precisely why I couldn't follow Steve's pointing, why I've never been able to tell where anyone is pointing. Researchers have found that people with amblyopia have a distorted sense of space, one that is laterally compressed and horizontally stretched. That means I should look higher and closer than I usually would.[1]

"Do you see the skinny tree that's coming out of the stump?" Sara says.

Two things about birders: We are not very good at tree identification, and we will work very hard to help other people see a particular bird. If you're the person everyone is helping, sometimes you must lie. If you're too honest, you'll all die of starvation or exposure.

"Oh, wait! I have an idea!" I say eventually. I pull out my cell phone and snap a picture of the trees. "Can you circle where the bird is?"

"I could," Steve says. "But it flew."

"I *hate* it when they fly," Sara announces.

"You hate it when birds fly?"

"I mean... You know what I mean," Sara says, laughing.

For my first trip to The Levi Lab, I stayed with my friend Heather — and I was an awful houseguest. I offered nothing: no conversation, no help with dinner, no entertaining her adorable daughter. For four days, you could find me either taking impossible vision tests at UC Berkeley or lying in Heather's bed, moaning while eating all of her pain relievers.

After two months of Vivid Vision, I'm returning to UC Berkeley for post-training tests. I don't even tell Heather that I'm going to be in town this time — rude, I know. But after three hours of migraine-inducing tests every day, I do not expect to be up for socializing. Indeed, my plan is to spend all of my free time lying on a hotel bed with my eyes closed, feeling sorry for myself. This time, I bring my own Tylenol.

Before heading to my first day of testing, I drink a bubble tea to steady my nerves, and then I buy another one to keep me company. Ding meets me at a back entrance to Minor Hall, and we head upstairs.

"You made a big improvement," he says. "Did you notice any differences?"

"No," I say. I keep hoping to see the world take on a new dimension, but it looks the same as always. "Well, I think I might have gotten a little better at threading a needle."

Hilary Lu sits me down at the computer, and I see that I'm going to be taking the exact same tests as before.

"Can I listen to an audiobook while I do the test?" I ask. I'm pretty sure that listening doesn't affect my vision, and it will make the next few days a lot less tedious.

"Sure," she says. "That's a good idea."

As I listen to *The Diana Chronicles*, I set my chin on the little plastic shelf and look at the computer screen. I expect to see either a horizontal line or a vertical one, but no—floating before my face, like it's no big deal, is the plus sign. My brain is listening to both my eyes *at the same time!*

I press the space bar, and the monochrome beach balls appear. The ones on top are clearly closer than the bottom ones, so I press the up arrow. The computer makes the "You got it correct" bleep. I feel gathering excitement in my chest. I take a breath and try to stay relaxed. *Bleep.* I got it right again! *Bleep.* And again!

Meanwhile, Princess Diana is finding that she's not so dumb after all. She's dating a famous Pakistani heart surgeon and has taken it upon herself to learn about cardiology and Islamic art. So many people told her that she was stupid just because she didn't do well in school. My heart swells with hope for her future, even though I know better.

The balls get smaller, and the apparent space between the near and the far ones diminishes, but I keep doing well—way better than chance.

Forty minutes pass without a whisper of a migraine. I'm surprisingly content, sitting in the dark with Princess Diana and my bubble tea, and the no-longer-impossible vision tests.

The computer screen fills with gobbledygook. "Done!" I shout, while skimming the page for my score. Lu and Ding appear, and I point at the ".85." "Does that mean 85 percent correct?" I ask.

"Wow," says Ding, eyes wide. "That's better than me!"

My heart is a hot-air balloon that's dropped all its sandbags. As both a teacher's pet and a people pleaser, this is my ideal way to spend an afternoon.

The next two days of testing are essentially the same: every time I finish a test, Ding and Lu come rushing in and exclaim over my high scores.

Guess what my new stereoacuity is? Nope, lower. Lower still. Divide that by half.

You got it: 179 arc seconds. As you may recall, I started at 3,207.

It's a vast improvement, but still far from normal.

I can't wait to tell Nasr and get him to rescan my brain.

In 2012, neuroscientist Bruce Bridgeman walked out of a movie theater and into a whole new world. "I was astonished to see a lamppost jump out from the background. Trees, cars, even people appeared in relief more vivid than I had ever experienced," he wrote.[2]

Bridgeman's path to stereovision was not intensive vision therapy, like Susan Barry received. He merely watched *Hugo* in 3D. Immersing himself in a world with exaggerated stereo cues—and doing it for a solid two hours—seemed to have "woken up" his binocular neurons. Amazingly, they stayed awake.

This was quite a surprise to Bridgeman, because as a vision neuroscientist, he knew about H&W's research. Bridgeman had been born with misaligned eyes, and he assumed he'd long since missed this window of opportunity.

Overall, he didn't think of himself as disabled—though he wasn't the best driver or bird-watcher.

"When we'd go out and people would look up and start

discussing some bird in the tree, I would still be looking for the bird when they were finished," he told the BBC. "For everybody else, the bird jumped out. But to me, it was just part of the background."[3]

Did his driving improve? Or his bird-watching? When I went to call Bridgeman and ask, I discovered that he died tragically in 2016, when he was hit by a bus in Taipei.

We do, at least, know that his 3D vision greatly enhanced his life. As he wrote in an email to Oliver Sacks, "I feel myself to be in the visual world rather than at it."[4] Reading this causes my head to tingle.

I often feel like more of an observer than a participant. Is this because I see the world the same way most people see a painting? Could learning to see stereoscopically allow me to feel, for the first time, truly immersed in my own life?

I want to try the Bridgeman technique for learning stereovision, but the only 3D movie currently in theaters is *Avatar: The Way of Water*. I did not see the first *Avatar*, because I don't really like movies — or TV shows for that matter. I used to think that my preference for reading was due to my intellectual heft, but now I see that my faceblindness is a big part of the story. It's hard to follow plots when you can't keep track of who's who.

Avatar 2 is three hours long, and I haven't had lunch, so I've smuggled in a Chipotle burrito. But since 3D movies have made me nauseated in the past, I've also brought a ginger ale and a barf bag. I've tucked all these supplies, quite conspicuously, into my giant coat. Reserving a seat in the dead center of the theater may have been a mistake.

Once I get settled in, the pre-movie commercials are already rolling. I see a dancing soda, and it doesn't look particularly plump.

"Does that look 3D to you?" I ask a man sitting two seats away from me.

"No," he says. "The commercials aren't in 3D, just the movie."

"Great," I say.

I don't know what I'm expecting today—something subtle, perhaps. That is not what I get. When *Avatar 2* begins, giant pointy rainforest plants are *in my face*. I flinch so violently, my ginger ale falls off my lap and rolls into the darkness underneath someone else's seat.

I'd like to retrieve my soda, but I can't take my eyes off the screen. Weird animals are screeching and flaring their crests. A mechanical foot slams down from above. My heart jumps. I'm going to get squashed!

Sadie, this is a movie, I tell myself. *Chill out.*

A blue man runs through the rainforest and takes a flying leap into an abyss. I hold my breath as he falls and falls through slanting light. My stomach lurches—but in a scared way, not a "Grab the barf bag" way.

I'm *terrified* of heights, but at the moment, I'm not feeling the usual fear. What I'm feeling is more fun, like I'm on a roller coaster.

Right after the movie, I happen to have a phone call scheduled with the legendary "Stereo Sue" Barry. I tell her that for the first time in my life I watched an action movie and it made me feel *excited*. Usually, my response to action sequences is similar to how I felt the first (and only) time I saw my brother wrestle. I don't understand what's happening, but I don't like it, and I'm afraid that someone is going to get hurt.

"Now I get why people watch these things," I say.

Barry says the same thing happened for her. Before she gained stereovision, she didn't understand the appeal of action movies.

"I would fall asleep during the action part of movies," she says.

"And so my kids would be like, 'I can't believe it—just as the space station is blowing up, Mom's asleep!'"

Barry suspects her reaction was the result of overstimulation. Fast-paced scenes require intensive visual processing, so action movies are quite effortful for people who are visually impaired.

I hadn't really put this together before—the idea that visual processing is harder work for me than it is for other people. But it makes so much sense.

"Is this why I'm so stressed out by clutter? Because I literally can't make sense of objects that are too jumbled up and close together?"

"Yeah," says Barry. "And are you more... Do you think you're more comfortable in the dark than other people?"

How did she know? I don't mention this, because it's too weird, but I love showering in the dark. I turn out all the lights, close the door, and put a towel down to keep light from leaking in. I find the soap by feel and sniff bottles of shampoo and conditioner to figure out which is which. Barry figured this out about herself when she noticed that, unlike her family members, she didn't feel the need to turn on the light every time she went up or down the stairs.

I wonder if stereoblind people compensate with other senses, like fully blind people do. Seeing isn't my forte, but I'm quite good at hearing. I can identify subtle differences between birdcalls, and I can play almost anything by ear. I consider asking Barry if she's good at these things too, but I'm running out of time—and what I really want to ask her is sort of personal.

"So, I've always had this feeling that I'm an observer, like I'm not really *in* the world," I say. "I'm hoping that if I get 3D vision I might feel a little bit less like an outsider. And I was just wondering if you had that experience."

Barry hasn't really considered this before, but she gamely talks through her thoughts. In a physical way, she says, she does feel more immersed—for instance, during a snowfall. "Then the question is, does that also overlap with how I feel about myself among others?" she asks.

"Yeah, like at a party or something."

"You know what, I think it does," she says. "I became much less shy as a result of stereovision, and much less embarrassed. Some of that might just be getting older. But yeah, I think I *am* more comfortable around people than I used to be."

A few weeks later, I have a video call with Nasr. Even though our meeting is clearly titled "Shahin Marvels Over Sadie's Newfound Stereovision," he refuses to stick to the agenda. My stereoacuity is still far from normal, he says.

"We could rescan you, but I can tell you now that we won't see anything," Nasr says. "Let me know when you get under 100 arc seconds."

I'm disappointed, but not devastated. Indeed, I'm grateful to amblyopia for giving me a rare glimpse into just how subjective vision is. Anytime I want, I can switch which eye I'm using and watch my perspective shift. After extensive training, I can also make myself see double, or view both images in a stereoscope at the same time. It's quite a trip to make things fade into or out of existence through nothing but the power of my own mind—a constant reminder that the visual world is both profoundly real and inescapably illusory.

After two years on the case, I think I have a pretty good handle on all the ways in which I am not quite neurotypical—but I'm about to find out that I'm wrong. Because there's another way of seeing, one that I didn't even realize was possible.

See ya, stereoblindness. We're off to see the wizard of aphantasia.

12

WE'RE ALL MAKING THE SAME MISTAKE

BEING FRIENDS WITH NINETEENTH-CENTURY POLYMATH SIR Francis Galton must have been exhausting. I imagine him walking through bustling London, circa 1880, causing everyone in his path to hide and scatter.

GALTON, *waving freshly printed papers:* Ah, Marshall, my old chum. Tally ho!
(*Anatomist John Marshall ducks behind a stagecoach.*)
GALTON: Marshall, do hold up for a minute.
MARSHALL, *feigning surprise:* Galton! Is that you, my good man? Forgive my lack of observation, but I must dash! (*Galton bids Marshall goodbye and shifts his attention to Charles Darwin, who has just exited a shop.*)
GALTON: Cousin! God save you!

DARWIN: Francis, how unexpected. (*He observes Galton's paperwork with a weary eye.*) You are not brandishing another one of your questionnaires, are you?

GALTON, *enthusiastically:* Indeed, I am, Charles. Freshly inked too! I could use your discerning eye...

DARWIN, *cutting him off, pointing to a cat crossing their path:* Oh, look, a rare specimen of *Felis catus*! I must follow it immediately! I'm afraid our chat must wait, Francis.

GALTON, *calling after him:* Does he have six toes? Do count them for me, dear cousin.

(*Scene ends with Galton standing alone, looking forlornly at his questionnaires.*)

After the publication of *On the Origin of Species* by his half cousin Charles Darwin, Galton dropped his faith in God and replaced it with a faith in science. He came to believe that it was our duty to understand and control nature—and his contribution would be to quantify everything in sight.

How many brushstrokes does it take to paint a masterpiece? How beautiful are the women of London as compared to those of Aberdeen? What is the perfect ratio of milk to tea? Galton's obsession with measurement wasn't pure silliness; along the way, he pioneered entire fields, including meteorology, statistics, and forensic science.

If only he'd stopped there. Toward the end of his life, Galton became obsessed with eugenics—the selective breeding of humans of "good" genetic stock. I'm sure I don't have to tell you what a deeply flawed concept that turned out to be. Interestingly, Galton himself never had children—possibly because he caught a sexually transmitted infection from a prostitute.[1]

WE'RE ALL MAKING THE SAME MISTAKE

Galton believed that the most desirable trait was intelligence—but he didn't know if intelligence was heritable. To find out, he wrote up a list of 40 open-ended questions, had them printed, and sent them to 170 gentlemen scientists—and in doing this, he invented the questionnaire. In addition to their detailed family pedigree, Galton asked his colleagues to assess their personality, character, religious fervor, business practices, talents, and education.

While interrogating the scientists of the day, Galton came across a curious phenomenon: some claimed to "see" numbers in their mind's eye. One man imagined numbers floating in the air, spaced evenly from left to right. Another scientist imagined them spiraling out from a central point. A third man reported that they appeared to him like layers of a cake, arranged in powers of ten.

Other friends of Galton said they had no idea what he was talking about.

Galton was astonished at the diversity of mental experiences reported by his men of science—and so he wrote another survey, the so-called Breakfast Table Questionnaire. It began with simple instructions. Bring up an image in your mind's eye of something you saw recently, such as this morning's breakfast table. Are the colors dim or vivid? Is it well illuminated or dim? Sharply defined or blurry? Is it larger or smaller than it is in reality? Is your field of vision larger or smaller than it is in real life?

Some of the respondents called the whole endeavor ridiculous. Anatomist John Marshall wrote, "I cannot place the slightest reliance on what a person says of his powers of seeing things with his mind's eye. The evidence is tainted at its very source. On the most mature reflection, I believe that the replies you will get will be full of fallacies; and not of scientific value."

Galton's gloss? "Marshall is obtuse," he wrote in a note to himself.[2]

Charles Darwin seemed to agree with Marshall, but he said so more politely: "I have answered the questions, as well as I could, but they are miserably answered, for I have never tried looking into my own mind. Unless others answer very much better than I can do, you will get no good from your queries."

Darwin also said that he was able to summon up a fairly vivid mental image of his breakfast. "Some objects [are] quite defined, a slice of cold beef, some grapes and a pear, the state of my plate when I had finished, and a few other objects are as distinct as if I had photos before me."[3]

Even so, Galton rated Darwin's imagery as medium-low. Indeed, his entire analysis seems to have been shaped by an a priori assumption that men of science, as a group, have poor powers of mental imagery. A 2006 reanalysis of his data found no significant differences in the visualization skills of scientists and nonscientists.[4]

While Galton's overarching conclusion hasn't quite held up, he was on to something. There are people who can't visualize *anything* in their mind's eye, and they do seem to have a slight advantage when it comes to abstract thought. These folks are somewhat over-represented in the STEM fields, but there aren't enough to pull down the average.

Galton made a leap of imagination that few have managed before or since. He realized that there's a hidden diversity of conscious experience in the masses of humankind. How tragic that a man with such a keen eye for differences would want to stamp them out.

My good friend Miriam is one of the first friends to visit me in West Virginia. A tiny, energetic part-time burlesque dancer,

WE'RE ALL MAKING THE SAME MISTAKE

Miriam is usually a ray of sunshine. Today, however, she's a gray storm cloud, heavy with rain.

"What's wrong?" I ask.

"I still can't stop thinking about Thad," she says.

I give her a big hug, but inwardly I'm groaning. Why is she still obsessing over this guy? He wasn't that great to begin with—and they broke up more than a year ago.

I've already given Miriam my patented three-step breakup advice. (1) Hide your ex's pictures. (2) Block him on social media. (3) Steer clear of him in real life. She's been following it, she swears, but it's just not working for her.

I'm confused—my cold-turkey method always works for me. Not only is it effective, it's also fast. Once, just a month after a rather dramatic breakup, I ran into my former live-in boyfriend on the street, and didn't feel a thing. I stopped to chat and remained cool as a cucumber while he had a visible panic attack. It couldn't have gone better.

I've learned the hard way that people do not want advice; they want you to listen and commiserate. So that's what I do. As I'm listening, I notice that Miriam's descriptions of her obsessive thoughts are positively cinematic.

"So, when you think about running into Thad, do you visualize it?" I ask.

"Yes, it's very visual. It's *more* than visual; it's immersive, like VR," Miriam says.

I ask for details, and Miriam sets up the scene like a movie director.

"We meet up in this particular café—it's underneath an Afghani place in Adams Morgan. It doesn't have booths, but I've given it booths, and we are sitting in one. The afternoon light is streaming through the glass entryway upstairs, giving the basement café a warm glow. I see Thad, on the other side of some old-timey table lamps. He's wearing a short-sleeved shirt that shows off his arms, and he smells earthy, like mulch..."

The specificity, the sensory details — if I had that kind of an imagination, I'd never leave the house!

Well, no wonder Miriam has trouble getting over breakups. It's impossible to avoid seeing someone when you have a full-scale 3D model of them in your head.

"It sounds like you have hyperphantasia," I say. "I think I have the opposite, aphantasia. I can't visualize at all."

I first read about aphantasia when my brother, Saul, sent me a link to a 2015 *New York Times* article that described the curious case of a sixty-five-year-old man called "MX."[5]

MX was referred to neurologist Adam Zeman after he complained of a problem that his doctor had never encountered before. Zeman had never encountered such a thing either. In an online video, MX — a sanguine man with a gentle Scottish accent — reprised his first conversation with Zeman: "Just a few weeks after I had my angioplasty... I noticed at night when I went to bed that I couldn't do what I usually did, which was, before going to sleep, thinking about my family, my children and my grandchildren, and picturing them. They just wouldn't come to mind."[6]

His backup technique for getting to sleep — counting down from ninety-nine and picturing each numeral — also didn't work. No matter what he tried to visualize, he simply found he couldn't do it anymore.

In addition to losing his visual memory, MX had lost his visual imagination. When he used to read books, he saw all the characters and settings in his mind's eye. Ever since his procedure, books had become merely words on a page.

Did something happen during the angioplasty? Zeman asked.

Yes, MX said. It hadn't alarmed him at the time, but he had noticed a "reverberation" in his head, and then some tingling in his left arm.[7]

This sounded like a small stroke to Zeman, but it didn't seem to have affected MX. He passed all the usual neurological tests with flying colors and—other than his loss of his visual imagination—MX seemed to be doing perfectly fine.

"It hasn't had any effect at all. My memory is as good as it was," MX said. While he couldn't imagine visiting his favorite places anymore, he was able to recall specific details of scenes and buildings. "I could remember them but I couldn't visualize them."[8]

Even though there was nothing to be done, Zeman wanted to further investigate MX's strange complaint for the sake of science, and MX was happy to help.

Zeman and his colleagues threw more tests at MX—asking him questions that would seem to require a visual imagination to answer. Which letters of the alphabet extend below the line? What's greener, an avocado or grass? MX answered without hesitation.

MX also passed a mental rotation test, but his pattern of scores was unusual.

Here, you try: As quickly as possible, say whether the following person is holding the flag in his left or right hand, and get a friend to clock you.

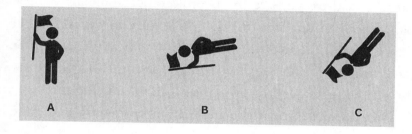

Normally, people take longer when the image has been rotated further from the person's normal, upright position. So, you should have taken the shortest amount of time for person A, who is oriented the way you'd expect; a little longer for person B, who has been rotated 99 degrees to the left; and the longest for person C, who has been rotated 140 degrees.

The fact that most people take longer with larger rotations suggests that people are actually turning the image in their mind like the hands on a clock.

Oddly, MX took the same amount of time regardless of the rotation, implying a mysterious, nonvisual technique.

Zeman also put MX in a scanner, showed him famous faces, and then asked him to visualize famous faces. While looking at the famous faces, MX's FFA activated as expected. But when he tried to imagine the faces, his FFA refused to light up. (Control subjects' FFAs lit up while looking at famous faces and imagining them.) Additionally, MX's prefrontal cortex was more activated than normal—perhaps because his conscious mind was working overtime, trying to get his visual areas to produce an image.[9]

After Zeman published a paper on aphantasia in 2010, he started hearing from people who claimed they had the same problem as MX—but, unlike MX, they'd *never* been able to visualize.

Here's a full transcription of my reaction to this article when I first read it: "Huh."

Usually I'd be more curious, but I was in the middle of my learning-to-drive-and-see-in-3D project, not to mention unpacking. My long to-do list wasn't the whole story, however. I was also grappling with the fact that my world was radically flattened compared to most people's—the possibility that my *imagination* was also seriously impoverished was too much to contemplate.

Now it seems I'm up for it, and Miriam's romantic predicament ignites my curiosity. I reread the article, and I am once again confronted with the fact that almost everyone besides me can conjure up all sorts of images in their mind's eye. Things that I thought were just figures of speech—daydreaming, imaginary friends, undressing someone with your eyes, counting sheep—are much more real than I realized.

Why didn't anyone tell me?

Now that I think about it, there were clues. I've been instructed, for instance, to deal with stage fright by imagining the audience naked. I thought I was doing it by thinking about how the audience *could* be naked. Meanwhile, my fellow musicians were gazing upon rows of bare torsos, and who knows what else. (People dancing! Body parts jiggling! This is starting to seem like terrible advice.)

I look up and notice Steve sitting on the couch, analyzing data coming from temperature probes in the brisket he's cooking. If anyone lives in a world of pure abstraction, it's him.

"Can you play mental movies in your head?" I ask.

"Yeah, but they aren't as vivid as Miriam's," he says.

This makes no sense at all. Of the two of us, I'm the creative one—the person who dresses in bright colors and entertains children with shadow puppets. How is it that Steve has a better imagination than me?

I spend the rest of Miriam's visit peppering her with questions, and boy, she has *a lot* going on in that tiny, adorable head of hers. She always has a commentary track or two running in her mind, planning what she's going to say and imagining how other people might react. Again, I had no idea this was an option. It does make sense, though. I'm often told to think before I speak, and now I realize that's something that most people can actually do.

While I can only think about one thing at a time, Miriam can simultaneously participate in the present while imagining moments from her past or visualizing the future. She can pause the tape, rewind it, and even try different courses of action. Sometimes, just for fun, she puts everyone in different outfits. "It's like playing with dolls," she says.

"Isn't it all terribly distracting?" I ask.

Miriam, who has ADHD, says it *is* distracting. Her busy mental life sometimes makes it hard for her to function.

Suddenly, I have a lot more sympathy for the fact that she is always running late.

Later that week, I call up my brother to talk about the aphantasia article he sent me all those years ago. After a quick chat about how neither of us can visualize, and apparently everyone else can, we have a minor disagreement.

"When you sent me the article..." I start to say.

"I didn't send it to you," he says. "You sent it to me."

"No way," I say. "You sent it to me."

We both search our emails and come up empty. Then Saul's wife, Katharine, chimes in with the answer. Saul heard about aphantasia on a podcast and then told me about it, and then I sent him the article.

Later that day, I call up Zeman—and when I tell him about this incident, he laughs. Apparently, not remembering your own life and then having to consult your spouse is a classic aphant move.

"About one-third of people with aphantasia also have poor autobiographical memories," he says. "But the relationship between memory and aphantasia is complex—there are some people with aphantasia who tell us they have the best memory in their family."

WE'RE ALL MAKING THE SAME MISTAKE

"Does aphantasia tend to run in families?" I ask. I've started polling mine, and I can't find any other aphants besides me and my brother.

Zeman says that it does, but there are probably a lot of genes involved, and no one has discovered any of them yet.

I mention all my other neurological quirks, and Zeman says that about a third of people with aphantasia also have prosopagnosia. However, the relationship between these conditions is unclear because, confusingly, there are plenty of aphants who are normal or even excellent at recognizing faces.

"My brother is one of those," I say.

"Aphants must have some kind of internal, visual representations—they just can't seem to use those representations to generate images," he says.

"Maybe there are multiple kinds of aphants," I say. "Maybe some of us really can't visualize at all, while others are visualizing but they don't realize it, because it's beneath the level of their conscious awareness."

Zeman doesn't know, but there probably are subgroups. In fact, a few have already emerged. He initially coined "aphantasia" to describe people with no visual imagination, but it's since also been applied to people who lack an aural imagination—people without a mind's ear. If you can't conjure up mental experiences in any sense modality, you are a "total aphant"—and that's the group my brother belongs in. I, in comparison, almost always have a song stuck in my head.

Suddenly, the concept of mental imagery makes a lot more sense to me. Inner seeing is like inner hearing! It's not as vivid or as loud as the real thing, but it's real enough to be annoying, distracting, or even entertaining.

I have a zillion more questions for Zeman: How many people can imagine tastes and smells? Is visualizing easier with your eyes open or closed? Do visualization-based approaches to therapy or coaching work for aphants? What should we be doing instead?

Zeman cuts me off. "No one knows. This research is in its infancy," he says. "Not much has been done in the hundred years since Galton."

Why? Galton's friend John Marshall may have said it best: "I cannot place the slightest reliance on what a person says of his powers of seeing things with his mind's eye. The evidence is tainted at its very source."

Objectively studying subjectivity is difficult, Zeman says, but it's not impossible. Thanks, in part, to Zeman's discovery of aphantasia and the flurry of interest it sparked, scientists are coming up with new and clever ways to validate people's self-reports — either by looking directly at brain activity or by measuring physiological responses that can't be consciously controlled.

Forget outer space. Inner space, it seems, is the final frontier.

For July 4th weekend, I fly to Minnesota to visit one of my oldest friends, Sybil. We're on a hike, supposedly to a waterfall, but the trail has narrowed to near invisibility, and it appears that we took a wrong turn. I'd whine, but Sybil is suffering more than me. Mosquitos the size of small aircraft are landing on my friend's exposed shoulders and leaving angry red welts behind.

I slap one of her attackers, and Sybil yelps in surprise.

"You're welcome," I say.

On the long schlep back to Sybil's car, it occurs to me that I don't know a thing about my best friend's inner life. In the approximately eight zillion hours we have spent together, it somehow has never come up. So, I ask her about it.

"Yeah, I can visualize things. Mostly just still pictures, but there are some moments I can replay like a video," she says.

Those tend to be highly emotional experiences, like when she gave birth to her daughter. Sybil proceeds to describe that day in cinematic detail—all the way down to the color, texture, and trajectory of her barf when it splattered against a yellow wall.

"Stop, please!" I beg.

"Wait a minute," Sybil says, whirling around to face me. "Since when do you care about this kind of thing?"

I make a confused sound.

"You used to argue with people, all the time, about how free will is an illusion, and how consciousness is an epiphenomenon," Sybil says.

I don't remember this specifically, but it's probably true. In my early twenties, I thought that debating metaphysics counted as flirting. (This may be yet another reason no one ever asked me out.)

While I was annoying, I was also (probably) right. In the last twenty years, the pile of evidence against free will has only grown. For instance, in a 2008 study, German researchers asked participants in fMRI machines to decide whether to use their right or left hand to tap a button. They could see a timer, and they were asked to note the moment when they made their decision. By looking at brain activation in the participants' premotor cortex, the scientists found they could predict which finger the participants were going to use up to *seven seconds* before the participants thought they'd decided.[10]

I've come to suspect that the conscious parts of our minds act like White House press secretaries. Even if they aren't in the loop, they pretend to be—and they're great at making up plausible reasons for our actions.

A case in point: In a 2005 study, Swedish researchers asked students to pick which of two photos featured the more attractive person. After the students chose, the researchers used sleight of hand to switch the pictures and then asked the students to explain their choice. The students happily pointed out particularly attractive traits from the photo *they had not actually chosen*.[11]

Later that day, Sybil summons up another old, embarrassing memory involving me. We're at a dinner party, and I'm chopping cilantro in the kitchen when I hear my name.

"Sadie was the best student in her rat-training class, in college," Sybil begins.

"No," I shout. "Don't tell this story."

"Now she has to tell it," our host correctly points out.

The name of the class was Research Methods in Psychology, and my rat, Templeton, deserves at least half the credit for our outstanding performance. Using the principle of positive reinforcement (rewarding a creature for a behavior you want it to repeat), I trained him to press a lever, and he caught on faster than all the other rats in our lab.

At first, I gave him a food pellet every time he pressed the lever, which resulted in a consistent, but not particularly fast, rate of lever pressing. To get him pressing that lever more quickly, I instituted what's known as a variable-ratio reinforcement schedule. I gave Templeton a treat every few presses — three on average, but the exact number varied. He started hitting that lever rapidly, barely pausing to enjoy his food pellet before getting back to work.

My professor was a behaviorist — a branch of psychology that had its heyday almost a century ago. It provided a simple answer to the long-standing question of how to objectively observe subjective experiences: Don't.

"Psychology as the behaviorist views it is a purely objective, experimental branch of natural science which needs introspection as little as do the sciences of chemistry and physics," wrote the movement's founder, John Watson.[12]

Subjective phenomena like inner speech and imagery are best left to philosophers—which is to say, crackpots, he argued.

Content warning: If you have been traumatized by applied behavioral analysis, you may want to skip to the next chapter.

When I encountered behaviorism, it was love at first sight.* Watson eloquently articulated something that I've always believed: thoughts are not really all that important. I mean, does anyone have any idea what they're thinking? I certainly don't.

I became truly obsessed with behaviorism after seeing how well it worked in the real world. After just twenty minutes of training, Templeton was hitting that lever like a pro. Within a few weeks, he was turning circles and standing on his hind legs on cue. We were quickly running out of tricks that could be performed in a psych lab, so I started looking around for other creatures to train.

My first target was my roommate, Cheryl. She was a perfectly lovely gal, except for her habit of playing the *Romeo + Juliet* soundtrack on repeat, all day, every day. After a few months of this, I found myself fantasizing about tossing Cheryl and/or her boom box out the window.

---

* Of course, there are many, many problems with treating humans (and other animals) like mindless automatons, one of which I am about to experience firsthand. For more information, see H. Kupferstein, "Evidence of Increased PTSD Symptoms in Autistics Exposed to Applied Behavior Analysis," *Advances in Autism* 4, no. 1 (2018): 19–29, https://doi.org/10.1108/AIA-08-2017-0016.

DO I KNOW YOU?

What did I do? I initiated a calm, mature conversation and we came to a mutually acceptable solution.

Ha ha, just kidding. What I actually did was apply a technique I'd learned in rat lab.

Extinguishing an established behavior—especially one with a built-in reward—is nearly impossible. Your only good option is to reward behaviors that are incompatible with the behavior you don't want to see. If you want your rat to stop pressing a lever, give it treats for doing something else, like turning circles. As for Cheryl, I immediately—without explanation—offered her candy whenever she played an album other than *Romeo + Juliet*.

Since I was new to experimental psychology at the time, I didn't keep good records, and I can't say for sure if she started playing the album less often. My defenestration fantasies abated, but there are many possible explanations for that. I may have accidentally trained *myself* to like Cheryl by giving her candy—a phenomenon called the Ben Franklin Effect. If you start being nice to someone, your feelings tend to follow suit.

My next lab rat was myself: For the first time in my life, I started going to the gym on a regular basis. I rarely missed a day—because if I did, my friend Melissa took away my network card, depriving me of internet access.

Buoyed by these successes, I turned my attention to my boyfriend, Neil.

Neil and I had been together all four years of high school, and we decided to stay together afterward, even though our colleges were more than a thousand miles apart. It was the '90s, so phones were still attached to walls and calling long-distance was expensive. My college's phone system charged even more than regular phone companies, so Neil got a calling card, and the plan was for him to call me.

This worked for about a semester, but then his calling dropped off precipitously. As I sat by the phone, waiting for it to ring, I became increasingly furious—and when Neil did call, I'd chew him out.

One day I had a realization: I was doing the equivalent of shocking Templeton for pressing the lever. No wonder Neil was calling me less and less often.

It was time to apply some scientific rigor to this problem. I stuck a chart by my bedside and started plotting Neil's calls. After establishing his baseline calling rate, I implemented a fixed-ratio reinforcement paradigm, whereby—instead of yelling at him—I aimed to be pleasant and entertaining. (Phone sex may have also been involved.)

It worked like a charm. Neil's rate of calling rocketed upward, and the slope of his graph almost perfectly matched the one I'd made of Templeton's lever pressing.

To get Neil to call even more often, I used a variable-ratio reinforcement schedule—only picking up every three calls on average. As expected, he started calling me like crazy. In addition to the joy of having my boyfriend back, my phone's constant ringing annoyed Cheryl—so it was a double win.

As I listen to Sybil recount my boyfriend training program, I wonder if my affinity with a theory that downplays the importance of thoughts and free will comes from the fact that I, personally, don't have any direct experience of my own mind at work.

And if this is true for me, what does it say about behaviorism's founder, John Watson?

As it turns out, I'm not the first person to wonder about Watson's inner life. In a 2009 paper, psychologist Bill Faw points out several passages in Watson's writings that suggest the famous psychologist had weak powers of imagination, both visual and auditory.[13]

For instance, in his 1913 article "Image and Affection in Behavior," Watson wrote:

> I may have to grant a few sporadic cases of imagery to him who will not be otherwise convinced, but I insist that the images of such a one are sporadic, and as unnecessary to his well-being and *well-thinking* as a few hairs more or less on his head.[14]

In the same paper, he cast doubt on the phenomena of inner monologues:

> It is implied in my words that there exists or ought to exist a method of observing implicit behavior [such as thought and images]. There is none at present. The larynx, I believe, is the seat of most of the phenomena.[15]

It seems that Watson (and I) made the classic mistake of assuming that our personal experiences are typical of all humans.

We're in good company. Some of Western philosophy's greatest thinkers have assumed that their version of consciousness was the norm. Socrates probably had a clear inner monologue, Faw writes, because he claimed that inner speech is synonymous with thinking. "Thought" Socrates defined as "the talk which the soul has with itself about any subjects which it considers…not with someone else, nor yet aloud, but in silence with oneself." Aristotle, on the other hand, must have been a visualizer, because he claimed, "The soul never thinks without an image."

WE'RE ALL MAKING THE SAME MISTAKE

Between Galton and Zeman, surprisingly few psychologists have entertained the idea that human consciousness may take different forms for different people. I do find a rare exception, however—a psychologist who has been studying this very question since the 1970s. I can't wait to give him a call.

13

QUANTIFYING QUIRKINESS

THE PROBLEM WITH INTROSPECTION AS A SCIENTIFIC METHOD isn't just that it's somewhere between hard and impossible to objectively verify. An even bigger issue is that thinking about your thoughts changes them. In the words of psychology's founder, William James, introspection is like "trying to turn up the gas quickly enough to see how the darkness looks."[1]

In the summer of 1973, while driving on the arrow-straight highways of the Midwest, Russell Hurlburt mulled over this very issue. He was heading to psychology graduate school at the University of South Dakota, and he wanted to study people's inner experiences — but he wasn't sure how to do it.

Hurlburt remembered a magazine article he'd recently read, which was about time management for busy executives. It said that if you want to know how you really spend your time, have your secretary pop in on you every so often and jot down what you're doing.

The problem of figuring out what an executive is up to struck Hurlburt as similar to figuring out what people's minds are up to. In both cases, retrospectively asking people to account for their activities is fraught with opportunities for errors. You need to collect accounts at random, throughout the day. In the case of people's inner lives, Hurlburt believed that the element of surprise would be key.

But how? You could hire secretaries to surprise participants at random intervals and write down the contents of their minds, but Hurlburt had a better idea. A former engineer, Hurlburt designed a device—in his mind, while driving—that would emit a high-pitched squeal every so often, which participants would respond to by taking notes on their inner experience right before the beep. A wearable black box, Hurlburt's invention looked like a pager—technology that hadn't yet been invented.

Hurlburt named his technique Descriptive Experience Sampling, or DES, and over the years, he's discovered five basic types of conscious experience: inner seeing, inner hearing, unsymbolized thinking,* feelings, and sensory awareness. There are other flavors of consciousness, Hurlburt says, and everyone seems to have their own personal recipe. Some people sample evenly from all the categories; other people divide their time between two or three. Then there are the rare purists, people who *always* think in pictures, for instance, and those who are simply awash in a sea of sensation or emotion. Adding to the complexity is the fact that

---

* Unsymbolized thoughts are cogitations that are not accompanied by imagery, inner speech, or anything else. Knowing that you need to grab your umbrella before leaving the house is one example.

QUANTIFYING QUIRKINESS

people can experience multiple modes of consciousness at the same time.

Hurlburt hates to generalize, but he has found some larger trends. For instance, in one study he found that students with inner monologues tend to be happier than those who think mostly in visions.[2] (I would have expected the opposite.) In another study, Hurlburt and his colleagues found that the inner lives of people with eating disorders tend to be extremely complex.[3]

In the eating disorders study, one participant with bulimia, "Jessica," happened to be watching the TV show *Scrubs* when her beeper beeped. In the show, a pretty, skinny woman had just entered a room, and all the men stopped and stared. A moment before the beep, Jessica had two inner monologues going at the same time — she said that one was emanating from the front of her head and one from the back. Here is what they were saying if you put them together, with the front voice italicized: "Why is it that movies and TV shows always have *blond, skinny* girls for *guys* to *stare* at?" Simultaneously, Jessica's brain was rifling through memories of TV shows and movies and coming up with examples of this trope.

Some of the participants also reported distressing visualizations. "Samantha," for instance, was beeped after eating a piece of cheese. At that moment, she saw a vision of herself as inaccurately fat. She felt a heaviness in her upper body and arms, which she interpreted as guilt, and had the unvoiced thought, *I could have done without that piece of cheese*. Overlaid atop the whole experience was an inchoate feeling of sadness.

I can't even imagine feeling and thinking so many things at the same time. And while correlation is not causation, I do wonder: Is it possible that my quiet inner life — and in particular, my *lack* of visualization — has kept me from being critical of my body?

This is a gift horse whose mouth I don't usually stare into, but I am notably unconcerned about how I look—and I grew up fully immersed in America's national pastime of fat-shaming. I watched every episode of *America's Next Top Model* and (unfortunately) a few episodes of *The Biggest Loser*. I watched Oprah roll out that wagon of fat, and I subscribed to several magazines that often plastered newly skinny women on their covers, celebrating their weight loss as if they'd won a Nobel Prize.

And yet, I largely avoided internalizing the poisonous beauty standards, standards that continue to haunt many of my friends. To this day, it often doesn't even occur to me to think about how I look to other people.

I wonder if this is aphantasia related. I have no visual memory or visual imagination, so if I am not currently looking in a mirror, I have no idea what I look like. And unless people are pointing and screaming at me, I default to assuming that I look basically fine—or possibly even *fine*.

I'm occasionally confronted with evidence that other people do look at me appraisingly, and they don't always like what they see. When I'm by a pool or at the beach, for instance, strangers will sometimes come up to me and say that I am "an inspiration." It took me a while before I realized they were implying that being a fat lady wearing a bathing suit in public makes me a brave body-confidence role model. I never ask for clarification, though. I choose to believe that they are all fans of my work.

I've also had the experience of random people applauding me for exercising. One time an entire construction crew cheered me on as I jogged up a hill. I thought that I looked like all the other runners in my neighborhood—sinewy, gazelle-like, and only lightly flushed. But after the unexpected approbation, I checked out my reflection and saw a woman so red-faced and sweaty, she appeared to be in dire need of medical attention. People. If you see a jogger on the verge of a cardiac event, don't applaud; call 911.

This realization of a potential bright side of aphantasia helps counter the mounting jealousy I've been feeling lately. I love reading — but I bet I'd love it even more if my brain turned everything I read into a movie (starring Cillian Murphy). I'm often lonely, but maybe I'd be able to chase that off if I could summon up visions of the people I love. Hell — if I could remember where I put my keys, that would be a major improvement in my life.

It might be possible to learn to visualize — I've reached out to a few potential teachers. But now I'm worried about what might happen if I succeed.

I schedule a video call with Hurlburt, with the hopes of talking him into sending me a beeper so I can try his method of capturing inner experiences, DES. As someone who is increasingly aware of what's *not* going on in my mind, I'd like to find out what *is*. I mean, it must be doing something, right?

Hurlburt appears on my screen, bespectacled, gray-haired, and impassive — the most professorial professor I've met to date.

I want to convince him that my brain is unusual enough to be worth studying, so I mention my prosopagnosia, stereoblindness, and probable aphantasia. Hurlburt is unimpressed.

"Most people are quite confident about the characteristics of their experience, but that doesn't make them right about it," Hurlburt says.

The professor delivers this bombshell as if it's an emotionally neutral fact, and not him telling me that I don't know my own dang mind. I'm offended — so I cover it up with a joke.

"When you go and speak at conferences, do you just open with 'None of you know what you're thinking or what anyone else is thinking'?" I ask, laughing.

"Maybe I don't open with it, but it's going to come out pretty soon," Hurlburt says.

To prove his point, Hurlburt offers to send me a beeper (yay!) and asks me to predict what I think we'll discover.

"I'm always [mentally] singing," I say. "I probably have a lot of feelings. Probably some unsymbolized thinking. But I don't think I have much inner monologue, and I don't think I have any visualization."

I might not know much, but I am pretty sure that I know what it's like to be me.

The first day I'm supposed to wear my beeper is a Thursday, but I forget until about 7 p.m. — right before I have a Zoom call with a DC teenager, to help with her homework. I layer a pair of headphones over my beeper earpiece, and I explain to Kendra that I might have to step away for a second to record my thoughts. She takes this in stride.

When my friend Dave talked me into joining this volunteer program, he waved away my concerns about the fact that I was never an outstanding student. All the way up to ninth grade, I got Cs and Ds. I started doing significantly better in high school, and I really hit my stride in college.

How did I turn this around? No one knows — and this mystery haunts me to this day. My anxiety dreams involve being sent back to grade school and failing every subject. This is an accurate assessment of what would really happen. I still cannot label states or countries on a map, spell, or do mental math. Do you remember the show *Are You Smarter than a 5th Grader?* I can tell you right now, any ten-year-old would flatten me. It wouldn't even be close.

This is why, most nights, poor Kendra has to figure things out for herself. Tonight, however, she has brought a writing assignment. I am so excited for the opportunity to prove that I am not totally useless. Writing is one of the two things I can do.*

Kendra's assignment is to write an essay comparing the election of 2020 to the election of 1876. To begin, we read newspaper articles about each election. Then we brainstorm similarities and differences.

"Time to write!" I say, opening up a Google Doc.

Kendra just sits there, watching the cursor blink.

I want to offer some advice, but I come up empty. How *does* one write? For me, it's a lot like digestion: The night before a story is due, I skim my notes, highlight quotes from interviews, and then go to bed. The next morning, I wake up, drink some coffee, and poop out an article. It's as simple as that.

"Just type whatever comes to mind," I say. "Pretend that you're telling someone a story."

Kendra continues to stare at the page, and I'm getting the sense that she wants me to write the paper for her—and I do too. It would be fun!

"Okay, just tell me what you're thinking," I prompt. Kendra replies, but unfortunately, so does Steve, who thinks I am talking to him. Steve is wondering what's in the fridge—an appliance that, I should note, is equally accessible to both of us.

Then the beeper beeps. If I were a computer, I'd be displaying a pinwheel or a little hourglass "thinking" icon.

"Sorry, one second," I say. "I gotta write something down."

---

* The other one, as you know, is lying still for long periods of time.

The next day, on my video chat with Hurlburt and his colleague Alek Krumm, I proudly trot out my first observation.

"Thursday at 7:10 p.m.... before the beep. I was making an annoyed face at Steve, my husband, who was next to me in bed. And I was pulling my lip down. And I was feeling my annoyed face and... the emotion of annoyance."

I pause for applause. The precision, the detail—I'm going to be the best DES participant ever.

Krumm's pretty, angular face is blank; Hurlburt looks thoughtful. They both have questions. Did I cognitively know that I was making a face, or did I feel the tug of the muscles of my face and infer that I was grimacing? Did the emotion and the physical sensations occur one after another, or simultaneously? What percent of my conscious experience did each of these experiences take up—was it 70-30, or more like 50-50?

And most importantly: Am I absolutely positive that the feeling of annoyance happened *before* the beep?

"Yeah, but the beep really added to it," I say.

Yesterday's irritation pales in comparison with what I am currently feeling. Hurlburt and Krumm are asking me to remember a random moment from my life with a level of care and precision generally reserved for defusing bombs.

When I figure out a way to mention this without sounding rude, Hurlburt explains that reporting on the contents of your own mind is much more difficult than many people realize.

"Some people, when they're angry, they say, 'I was seeing red,' and they mean that absolutely descriptively," as if they put on red-tinted glasses, Hurlburt says. "Other people who say, 'I was seeing red,' mean it entirely metaphorically. There's no red feature to their experience at all."

QUANTIFYING QUIRKINESS

I hate to admit it, but he's right. I went forty-odd years without understanding what "counting sheep" means to most people. Words conceal as much as they reveal.

"By the way," Hurlburt says, "we generally just throw out the first day [of DES analysis] because there're too many possibilities for mistakes."

"Right. Of course," I say, my eyes growing as big as saucers.

We continue for seven more weeks: I wear Hurlburt's beeper for a few hours each week, and I write down my inner experiences when it beeps. I notice myself getting better at catching my mind in the act of thinking. The trick is to write down the contents of my consciousness *immediately*, and to include every detail. If I get lazy and skip something, there's no chance I'll remember it later.

We end up capturing and analyzing a total of thirty-nine discrete moments from my life. But before we're done, the three of us go through all of them again, just to make sure that we're in agreement. Finally, they call me with my results.

My most common conscious experience is nothing. About a sixth of the time, my mind is empty—and that's quite unusual, Hurlburt says. But he and Krumm are not content to leave it at that. Apparently, I experience six different variations on "nothing," plotted across two different dimensions: Do I know *that* I am thinking, and do I know what I am thinking *about*? This results in a variety of experiences, from the specific ("I'm thinking that my youngest brother looks mature in his Army uniform") to the vague ("I know I'm thinking about something, but I don't know what") to the specifically vague ("I know that I'm thinking about nothing") to the Zen ("I don't know what I'm thinking about or even that I'm thinking at all").

My second-most-common conscious experience is sensory awareness. For instance, I'm noticing the sharp contrast between typed text and white paper, or I'm taking in the bright whiteness of a breadcrumb.

And my third is words—usually unvoiced individual words or even letters, nothing at all like Miriam's inner monologue.

Overall, I was right! I *did* know what was going on in my mind: a lot of nothing. This makes me feel oddly triumphant. I may have an empty head, but hey, at least I'm aware of it. That counts as sentience—right?

On the other hand, I incorrectly predicted that I'd be awash in big emotions. I think I made this mistake because when I do feel emotions, they come out of nowhere and overwhelm me. Thankfully, this doesn't happen very often.

Krumm and Hurlburt are working on writing an entire paper, just about me! Here's what they have to say about my unusually blank mind:

> Sadie's directly apprehended experience of thinking without the content of that thinking being apprehended is indeed unusual in our participants. Nearly everyone who experiences thinking experiences the about what of that thinking—indeed, the about what is often (probably usually) more salient than the how or the fact of thinking. Similarly, the word-by-word or letter-by-letter experience of semantic content is very unusual in DES participants. In by far the majority of DES participants, words are handmaidens of meaning, not of particular interest in themselves. As far as experience is concerned, people generally speak (innerly or aloud) meaningful utterances, not strings of words. Sadie's experiences of words were frequently of

the individual words themselves, stripped or separated from their semantic role. Of course she was speaking (or, in many cases, reading) meaningful sentences constructed of skillfully strung together words, but her experience was of the words, not of the arc of meaning. That is quite unusual.

They suspect that my faceblindness and my (probable) aphantasia are both facets of a larger mind-blindness. "I" have no idea what the rest of my brain is up to.

Or, as they put it:

It seems fair to say that one of the dominant characteristics of Sadie's experience, a characteristic that is quite unusual, could be said to be her blindness to her own thought and emotional content: she often apprehends herself to be thinking without apprehending what that thinking is about; she often apprehends herself as producing words without directly apprehending the arc of meaning that connects those words. Might that be a superordinate characteristic to Sadie, who often experiences seeing a face without apprehending the person whose face it is?

That sounds about right to me. What *I* find surprising is the fact that it's not true for everyone.

While I've now participated in the most rigorous version of self-report imaginable, it is still self-report. I could be putting my hand on the scale, unconsciously attempting to gain entry into the cool kids' club.

The aphants *are* pretty nifty. I've been lurking on the aphantasia subreddit, and I'm impressed with how smart everyone is. These folks have taken it upon themselves to read and understand all the aphantasia papers, and their debates are unusually collegial and well-informed. An inability to visualize does seem to go hand in hand with a scientific mindset.

Perhaps this is why I'm determined to find objective evidence about whether I belong to the aphantasia crew.

If aphantasia is a cool kids' club, I'd like to nominate Craig Venter to be president. Known as the "bad boy of biology," he's famous for starting the company that beat his former employer, the National Institutes of Health, in the race to sequence the human genome. Since then, he's been tootling around on his yacht, collecting the DNA of oceanic microorganisms and trying to build carbon-eating synthetic life-forms that spit out biofuels.

I picked up his memoir, *A Life Decoded*, because I'd heard that he was a fellow aphant, and I was curious to see how he managed the unique challenges of being an aphant autobiographer.

Overall, his autobiography is very plot focused, and there are not many lush visual descriptions. There are a few, though. In Vietnam, Venter remembers seeing "bomb-scarred rice paddies, and the ruins of thatch-and-mud-wattle hamlets." There's even a specific instance where Venter claims to visualize a yacht he was designing. "I loved this particular project best when I had built its frame — when my imagination could fill in all the details of the completed boat... I still find that the final version often falls short of the one glimpsed in my head during construction."[4]

Was Venter visualizing or was he just being metaphorical? If he was being metaphorical, I do that too. In chapter 12, for instance, I imagined tossing my roommate out the window. However, I just thought about this conceptually, not in images.

Without any visual memory, writing lush descriptions is a challenge. Sometimes I cheat. When writing about watching

Avatar 2, I rewatched the first ten minutes and tried to remember what seeing it in the theater in 3D looked and felt like. I checked my memory against an unhinged voicemail I left Miriam right after seeing the movie — one that I saved, transcribed, and pasted into my journal. Despite my efforts, I suspect that this book, like Venter's, is short on sensory details but (hopefully) long on ideas.

Compared to mine, Venter's story is action-packed. After high school, he was drafted to fight in Vietnam, where he ended up running an emergency room at a Navy hospital in Da Nang. To keep sane, Venter would go swimming in the South China Sea. One afternoon, he felt something at his ankle. He reached down and grabbed the mystery object — and it turned out to be a venomous sea snake. If he let the snake go, Venter was sure it would bite him, so he killed it with his bare hands.

I didn't believe a word of this story until I found a photo of Venter standing on a beach in cutoff jorts, brandishing a limp and lengthy snake. Venter kept the skin. It still hangs in his office.

I'm not much further into the book when I realize that I have just twenty minutes before my video call with Venter. I skim a few newspaper articles critical of him — calling him an egomaniac who values profit over science — and then I'm plumb out of time. I open a new window on my laptop, and Venter appears. He's ruddy and hale, younger looking than his current age (seventy-six). Behind him is a shelf with model sailboats and a book by Ernest Hemingway. I can't make out the title, but *The Old Man and the Sea* seems like a good bet.

"When did you first realize that you couldn't visualize like other people?" I ask.

"It was after Vietnam," Venter says. He and his first wife, Barbara, were taking community college classes together. Barbara barely had to study for exams, due to her photographic memory.

Venter, in comparison, had a terrible memory for facts. He could only remember things if they fit into a larger context, if he could really integrate them into his understanding of the world.

"No way!" I say. "Me too!"

I cringe as I say this. Comparing myself to Venter is ridiculous. He's one of the greatest scientists of his generation. I am a moderately successful science writer. And yet, this is *exactly* my experience.

"How are you at spelling?" I ask.

"When I was in seventh grade, I refused to take spelling tests," he says. "I thought they were just the stupidest thing—you're given this list of words to memorize and the next day regurgitate them. As a result, I can't spell today—even though I worked out the human genetic code, the biggest spelling challenge in history," Venter says.

I don't tell Venter this, but I am an atrocious speller. I just misspelled "atrocious," and now I've misspelled "misspelled." Thank God for spelcheck.

Venter was a D student in high school. "I *barely* graduated," he says. He did much better in college for several reasons, but one big one was that the professors valued what he was good at: understanding and synthesizing concepts.

"I had a B average in college...but I never got a single B. I either had As or Cs. And I'd get As in the classes that required comprehensive understanding and Cs in the classes that were more based on rote memorization," he says. "But once I got into research, I found my conceptual thinking had me leap ahead of all my colleagues."

This story is extremely familiar to me. I was a D student until I hit high school. Then, suddenly, I became a B+ student. Talking to Venter is giving me some insight into why this happened.

QUANTIFYING QUIRKINESS

Grade school—at least the ones I went to—is all about memorizing facts and retrieving them on command, and I am almost comically bad at recalling information out of context. I have been kicked off of two different bar-trivia teams. Once, I walked up to a group of friends playing Scrabble, scrutinized the board, and caught an error.

"'Tuh-heh'? 'Tuh-heh' isn't a word!" I said.

"'The'?" Steve said.

Venter believes that the US school system privileges the visualizers, people whose brains naturally hold on to details. Venter once gave a commencement speech to a group of top high school graduates from across the country, and he polled them from the podium.

"I...asked how many of them had photographic memories. And it was like 98 percent of the audience. And there were only maybe a hundred hands or so out of the twenty thousand that thought they had aphantasia," he says. "That's why I've argued it would be great to have a straightforward either genetic test or...MRI diagnostic early on. And people with aphantasia should have a different education stream than one based on rote memorization."

Wouldn't *everyone* benefit from an education that privileges understanding over memorization? Also, wouldn't Venter's testing plan lead to a *Gattaca*-like situation, where DNA is destiny? These are both good questions that I try to ask, but not very hard.

Venter says that while aphantasia was a major disadvantage in grade school, his facility with big-picture thinking helped him distinguish himself as a scientist and a manager.

"The way my brain works has probably contributed more to my success than any other trait that I have," he says. "It's not just that you can't see pictures. It's that your whole way of looking at the world is different from most people."

14

TRIANGULATING THE TRUTH

Y OU KNOW THAT FEELING YOU GET AFTER GOING TO A TAPAS restaurant? You're full—stuffed, even—but still vaguely unsatisfied? That's how I felt about halfway through college. I'd taken every psychology class and they'd all sidestepped my big, unspoken question: What is it like to be someone else?

I got the sense that there was something wrong, perhaps even shameful, about my wish to understand other creatures' subjective experience, so I dropped it.

My midlife crisis has caused this question to come roaring back with renewed urgency. While it's true that everyone assumes all the other minds in the world are basically like their own, this is a bigger problem for those of us with unusual minds. Simply by dint of our own uniqueness, we're more likely to misunderstand and misinterpret other people's behavior.

For instance, because of my faceblindness, I've always believed that I'm a little famous. How else to explain all these folks who seem to know me, whom I have never met before in my life?

I also don't have the powers of visual memory that most other humans have. So when other people notice something about me has changed, that I've gained a few pounds or bought a new dress, I feel weirdly scrutinized. What are you, some kind of stalker?

My failure to understand all the hidden neurodiversity in the world has caused me to misinterpret some of my closest friends. Miriam isn't willfully obsessing over her past. In a way that I find difficult to imagine, her past is quite present. Sybil used to annoy me by worrying if she left the oven on—but I have more sympathy now that I realize her mind plays vivid movies of her house burning to the ground. Steve isn't choosing to "forget" to do the dishes. He really, truly forgets.

Are we all stuck in our own experiential bubbles, with only the flimsy filament of language to connect us? And what about animals—can we ever know what it's like to be, say, a bat?

According to the philosopher Thomas Nagel, the answer is no.

In his famous essay "What Is It Like to Be a Bat?" Nagel argues that it's impossible to imagine inhabiting the mind of another creature. You can learn about the sensory abilities of the bat. You can investigate how it perceives time, where it feels safe, and what activities it seems to enjoy. But no matter how much you know about bats, the very best you can do is to imagine what it would be like for *you* to be a bat. What it's like for *a bat* to be a bat—a bat that's been fully immersed in battyness since its batty birth—well, that's another issue entirely.

This is similar to how, even if I succeed in learning to see in 3D, I will never experience what the world looks like to people whose two eyes have been working as a team their entire lives.

TRIANGULATING THE TRUTH

And if you were to lose an eye, your world would not look as flat as mine, because your brain already knows how the 3D world looks, and it fills in those details.*

Despite this limitation, there's still a lot to be gained by trying to imagine other creatures' subjective experiences. Since we humans are such visual creatures, I think it makes sense to use visual analogies. Dogs, on the other hand, are smell focused. Smells linger in a way that light does not. A dog's world might be a little like a long-exposure photograph. They can "see" the afterimage of the hot dog that was dropped on the sidewalk yesterday. With a single intoxicating inhalation at the base of a popular pee tree, they can catch up on all the neighborhood gossip: who's pregnant, who's new in town, who's stressed out because they just got replaced by a human baby, etc.

Trying to understand the inner lives of others requires a great leap of imagination—and we will always fall short. But it's important to try. If we don't, we run the risk of putting dogs in expensive human clothes and calling them "pampered," when they really just want to be naked with their noses in something disgusting.

Humans are easier to understand, because we can describe our experiences in words—but everyone is still stranded in their own personal experience bubbles. It helps to pay attention to what kinds of things come easily to you as compared to others, and what people seem to expect you to be able to do that you can't. Popular culture can help: movies are made with the average person in mind, and one big clue that you might be faceblind is trouble following plots, because you can't keep track of who is who.

---

* When this happened to Oliver Sacks, his visual world gradually deflated during the months that followed. Sacks also told the story of a painter whose very conception of color slowly drained away in the years after he lost his color vision to a stroke.

If you start asking questions, and if you aren't afraid to get rather granular, you may be amazed at what you discover. Did you know that, for most people, left and right are as obvious as up and down?

If you were a neurotypical person beaming into my lived experience, here's what I think it would be like: You'd open your eyes and find yourself in a world that looks flat, like a painting. This world is populated with NPCs (non-player characters) because you can't tell people apart. There are only two types of cars, small and large, but they come in a variety of colors. Your mind is usually peaceful, though sometimes you feel a roiling unease beneath the surface. You have no idea what you're thinking about until you say it or write it down. If you can't figure something out, your only recourse is to go for a walk or a swim and hope your mind eventually coughs out an answer.

There's one major difference I have left out, because I don't know it yet. Suffice it to say, it's a biggie.

At the vanguard of the objective study of subjectivity is Joel Pearson's lab, at the University of New South Wales in Australia. In one study, Pearson and his colleagues recruited visualizers and non-visualizers and showed them two different videos: a slideshow of scary pictures of a shark attack and a slideshow that told a scary story of a shark attack in words, one phrase at a time. While this was happening, the researchers measured the participants' perspiration, an involuntary indicator of emotional arousal.

The researchers hypothesized that actually seeing blood in the water would be scarier than reading about it—and that was the case. Everyone—visualizers and non-visualizers alike—sweated like crazy while watching the scary movie. It was while reading the story that a difference emerged: the visualizers, who reported that they were seeing the attack in their mind's eye, got all sweaty again, while the non-visualizers stayed chill and dry.[1]

In a follow-up experiment, the scientists took advantage of a phenomenon known as binocular rivalry. The way it works is to use a stereoscope to briefly show two images at the same time — one per eye. A participant will only consciously register one of the images. Pearson and his students discovered that they could prime neurotypical people — making it more likely they would see one of the images — by asking them to imagine it first. This didn't work with the aphant participants.

Another clever experiment took advantage of the pupillary dilation response. Simply *imagining* bright patches of color caused visualizers' pupils to constrict as if they were looking into a bright light. The aphants, in contrast, had pupils that stayed the same size no matter what they tried to imagine.[2]

These experiments provide evidence that aphantasia is a real phenomenon, and not just an illusion caused by the difficulty of communicating one's inner experiences. I'd like to ask my pupils and sweat glands if they agree with *my* self-report, so I set up a video chat with Alexei Dawes, a postdoc at Pearson's lab.

With his laid-back demeanor and surfer's tan, Dawes confirms all my Australian stereotypes.

Dawes says that in addition to being interesting in their own right, aphants provide a natural "knockout" model for researchers who want to understand how visualization works.

The results of the shark-attack study, for instance, show that visualization links more strongly with emotion than words. Given this finding, you might expect that memories of visualizers are more emotionally evocative than memories of aphants. Additionally, you wouldn't expect aphants to get as swept up in books as visualizers do.

"Do you imagine the characters and scenes when you read?" Dawes asks.

"No!" I say. "I didn't know that was possible."

I never understood why people get so worked up about movie adaptations of their favorite books—but now I do! If you've already cast your favorite book, imagined the costumes, and pictured the settings, you're going to be mad if some director gets it wrong.

"You know the intro to *Reading Rainbow*?"

Dawes does not, so I describe it to him.

"Kids open books, and that transforms their world into magic castles and pirate ships," I say. Then I break into song: "Butterfly in the sky, I can fly twice as high. Just take a look, it's in a book, a Reading Rainbow."

I always thought that title sequence was ridiculous—a transparent attempt to trick dumb kids into reading. Now I'm realizing it wasn't a trick. For many people, reading really is a multisensory experience.

Before we hang up, I ask Dawes if they need any experiment participants. My goal, I explain, is to gain the deep scientific insight that one can only gain on an Australian vacation that also counts as a tax deduction.

Dawes says they aren't collecting any data at the moment, but he recommends I get in touch with Wilma Bainbridge, an aphantasia researcher at the University of Chicago. So, I do, and Bainbridge offers to peek into my head the old-fashioned way—with fMRI.

I'm so excited, I do all my afternoon chores while singing variations on the *Reading Rainbow* song. By the time I'm done folding my laundry, it has mutated into something like:

TRIANGULATING THE TRUTH

In the MRI / Impossible to lie
Take a peek / My brain's unique
It's aphantasia!

As you may remember from chapters 3 and 9, visual information starts at your eyeballs and then gets beamed to the very back of your skull for processing. From there, it progresses forward, in the general direction of your nose, through a series of areas creatively named V1, V2, V3, etc.

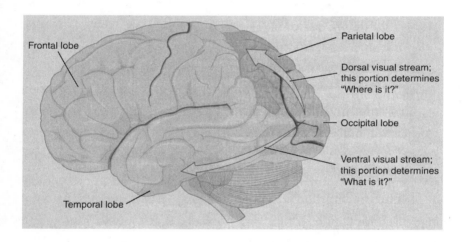

I left out a little twist: you actually have two visual-processing streams. They run in separate, parallel tracks until they hit the back of your skull. On the return trip toward your nose, one stream takes the high road, while the other one takes the low road. The high road, also known as the dorsal stream, captures *where* things are. The low road, known as the ventral stream, identifies *what* you are looking at.

The "where" system is evolutionarily older. It is colorblind, processes information very quickly, and doesn't capture much detail. If you need to catch a fly with your sticky tongue, the dorsal stream of vision is key. The "what" system is a more recent evolutionary innovation. It is slower, sees in color, and captures fine details. The "what" stream is all about object identification, and it includes our old friend the FFA.

When you visualize, it activates the same systems as when you actually see—but the process seems to work in reverse. So, visual processing goes from back to front, and visual imagination happens from front to back. In people with aphantasia, there's evidence that attempts to visualize activate the dorsal "where" pathway, but information on the ventral "what" pathway stalls out.

This is according to research by Bainbridge's lab. In a 2021 study, they asked aphants and neurotypical folks to look at a series of four photos and then to draw them from memory. The aphants drew fewer objects than the control group and used less color. They did, however, put the few things they remembered in the correct places.[3]

While the neurotypical people remembered more details, they were also more likely to add in objects that weren't there. The aphants were more accurate in this respect—they forgot most of what they saw, but when they remembered something, that memory was more likely to be accurate.

I wonder if this is also part of the aphant affinity with science: we believe it's better to overlook something than see something that's not there. (This is also true in journalism.)

I tried out Bainbridge's drawing task, and as you can see on the next page, I didn't remember many objects, and I labeled the window, sink, and cabinet with words. This is typical of people with aphantasia. "People with aphantasia love using words," Bainbridge says.

You'll also notice that I changed the centerpiece from a chef drinking wine to a chef drinking beer. That means, when I encoded the image in my mind, I jumped up a level of abstraction, from "wine" to "alcoholic beverage." Then, when I tried to translate my concept-based memory back into a concrete object, I picked the wrong alcoholic beverage.

I think we can all agree that my version is an improvement. An Italian chef drinking wine? So cliché!

I fly to Chicago and meet grad student Emma Megla in the grand lobby of what turns out to be a rather plain building. We exchange pleasantries as she ushers me upstairs and into a room that almost looks like a medical office, but not quite. In the corner is a fake MRI machine. People who are anxious about getting a brain scan practice with the dummy MRI before getting cozy with the big magnet.

"But we don't need to do that with you," she says. "You've had MRIs before, right?"

"Right," I say. I'm about to start bragging when I realize that I have made a rookie mistake. I'm wearing sequins. I pull at the shiny discs on my shirt.

"Are these metal or plastic?" I ask Megla. Tall and pale, with big round glasses, Megla radiates quiet intelligence, so I almost expect her to know.

"I'm sure we can find you a hospital gown," says Bainbridge, who has arrived a little late.

Megla opens her laptop and brings up a slideshow of photos. My job is to rate how familiar they look to me. Someone's messy room — not familiar. A smiling woman with kinky black hair — not familiar. My dad — familiar! My old apartment — familiar. My old neighborhood's Starbucks — very, very familiar.

"I forgot that I sent you these pictures," I say. "This is making me miss DC."

In the scanner, I watch the same slideshow, but my task is different this time: after each photo, I'm supposed to visualize what

I just saw—and I only get a few seconds to try. *Seconds!* Not only can most people hallucinate on demand, apparently you guys can do it instantaneously.

Later, over lunch in an airy faculty dining room, I ask Bainbridge and Megla if they think it might be possible for me to learn to visualize, and if there are any potential downsides. At least, I try to. I've had too much caffeine, so my question comes out as more of a monologue.

"I've found an optometrist who teaches people to visualize as part of his larger vision therapy program, and I have a video call with him next week, but now I'm worried..." I say, before sharing my theory of how aphantasia has kept me safe from harmful beauty standards. "I mean, I certainly apply them to other people, but since I can't see myself, I forget to care what I look like..."

Bainbridge and Megla are too polite to say so, but my current appearance supports my thesis. I forgot to pack a hairbrush, so my ponytail is in bird's nest territory. Bainbridge, in contrast, looks like she just stepped out of a Pantene commercial. Her hair is long, wavy, and impossibly glossy, and her dress, unlike mine, is free of wrinkles and food stains.

Bainbridge says that it's unlikely I'll end up haunted by unwanted visions.

"You'll probably have to work hard to see even a faint image," she says.

I decide to go for it—for science, and for the experience. I'd like to at least glimpse how the other 98 percent live.

"If I succeed, maybe I can come back and get scanned again?" I ask.

"Sure!" says Bainbridge. "I think we could do that."

* * *

DO I KNOW YOU?

The day after my visit to the Bainbridge's lab, I wake up early, acquire an enormous iced tea from Starbucks, and consider my options. I have five hours to kill before my 11 a.m. appointment downtown, so I decide to wander in that direction and see how far I get.

A half mile into my walk, I happen upon a diner festooned with signs proclaiming that it is Barack Obama's favorite breakfast spot. It looks like a basic greasy spoon to me, but who am I to argue with an ex-president?

After hiding my half-finished iced tea in my coat pocket, I head inside. Men in paper hats are standing behind a long sneeze guard, glopping food onto plastic plates. I order a plate of eggs, potatoes, and toast and sit down at a sticky table.

As I eat, I get out my little reporter's notebook and take notes: The floor, I write, is a dingy brown parquet. The people next to me seem to be dressed for church and they are arguing in a foreign language — Greek maybe. Above the big picture window are clocks for Athens, New York, Chicago, and London, and not one of them is giving the correct time for their respective city, or anywhere. It's almost impressive.

If someone were watching me, they might think I'm a restaurant reviewer, but this is just what I do when I'm bored. I take notes — describing things or people or scenes for my own amusement. I won't remember the scene, and I'll probably lose my notebook too, but the words I write down are likely to stick with me if I describe something well.

I get so engrossed in my note-taking, I lose track of time. (This happens to me a lot. I suspect I can add "timeblind" to my list.) When I look at my phone, I see that I've squandered my lead and am now running late. I stop by the bathroom on my way out—and I'm in such a rush, I do not check the state of the toilet seat before sitting down.

Oh, what a grievous blunder. My rear end lands in a pool of liquid—a puddle so deep, the impact of my butt sends it splashing in all directions. It's running down my pants legs and soaking my underwear—formerly white, now rusty brown. I am literally gagging. *I just sat in period pee.*

I'd like to toss my pants and underwear into the biohazard bin, but my shirt is not long enough to cover my butt. I'm going to have to stay in these gross clothes and hightail it back to the Airbnb to change.

At least the liquid isn't body-temperature warm. That would be worse. Come to think of it, it's colder than you'd expect—like, ice cold. That's weird...

My iced tea! I'd forgotten that I had hidden it in my coat pocket. When I sat down on the toilet, it must have tipped over and dumped tea everywhere. A quick investigation turns up confirmatory evidence: ice on the bathroom floor and a mostly empty Starbucks cup.

This is a huge relief; it puts me in a good mood for the rest of the day.

When I get into the Lyft, I tell my driver the entire story—and it occurs to me that whenever something terrible or strange happens to me, I tell someone as soon as possible. I don't have the mental scratch pad other people have, where they can just tell stories quietly to themselves. I have to say it out loud or write it down—and that process means that I must put all my experiences in the form of stories. They have to make logical sense, extraneous information in the setup must turn out to be relevant later, they must be interesting enough to keep someone's attention, and they have to hew to the classic setup-conflict-resolution arc.

According to the psychologist Dan McAdams, story-ifying your life is a great way to make sense of it. Without a narrative format, life is just a series of random, puzzling events. You're a sailor being battered by a cold and indifferent sea. He's found that people who see their lives this way tend to be depressed and ineffective. It's better to see yourself as the main character, the hero of your own story—someone who faces challenges and triumphs, or at least loses and learns. The indifferent universe story may be more accurate, strictly speaking, but humans need a reason to keep going; otherwise, we stagnate and despair.[4]

All people are storytellers, but I am more story focused than most. In college, my friends made me a flag that read, "Story Time." I wonder if this is because my memory is so poor, and if it's the source of my writing abilities.

While I yammer, the weather goes from "beautiful spring day" to "wintery mix." I bid my driver farewell, sprint through freezing rain into a nondescript office building, and follow the signs for a place called Float Sixty.

Float Sixty is a small chain of wellness centers with sensory deprivation tanks. Most people come to de-stress, but I'm here to kick off my visualization quest. My hope is that by floating in body-temperature water in the dark, my brain will get so bored, it will have no choice but to entertain me with imagery.

The odds are in my favor—most people who try out sensory deprivation have some kind of psychotic-spectrum experience, such as hearing voices or seeing things.[5] Joe Rogan, for instance, experienced a vivid hallucination that he was in a jungle listening to Indigenous people talking in their native language—and he suddenly realized that he could understand what they were saying.

"It was so extreme. I could smell the rain. I could feel the moisture in the air. I could see the leaves all around me. I could hear the sounds of the forest," he said.

I was a little disappointed when Rogan added that the hallucination wasn't just due to the tank—he was also, in his words, "tripping balls."

I'm not going to do drugs (other than caffeine) because involuntary imagery does not count as visualization. Lots of aphants can hallucinate on drugs, and almost all of us dream in images. Where we fall short is in intentionally visualizing—it's almost as if our prefrontal cortexes are trying to send commands backward to our visual areas, but the messages are not getting through. But if some other part of the brain asks for visuals, the visual areas are happy to oblige.

I'm led to a private room, where I shed my soggy clothes and slide into a pool of warm salt water. The light is on a motion sensor, so it should turn itself off soon.

My plan is to space out and see if any visions arise. If that doesn't work, I'll try a technique called image streaming, which involves describing, out loud, absolutely everything that I "see" in my mind's eye. To get started, I will lightly rub my eyes to make blobs of colors appear—and hopefully my brain will transform these Rorschach blots into something more interesting.

I breathe deeply and do a progressive relaxation exercise, a mental scan of my body. I start at my feet and then go to my ankles, and...I fall asleep. I wake up with a full-body twitch—a weird sensation that's amplified by saltwater waves ricocheting between my body and the sides of the tank. *Where am I? Why am I wet?* Oh, right. Now I remember.

This time, I focus my attention on what I "see" behind my eyelids. It's the usual field of black with green-red highlights. Then I notice a flicker of light green in just my right eye. I follow it as it jumps to the left, in a series of jerky movements. Then I drift off to sleep again.

When I wake up, I exhort myself to really focus. I open my eyes as wide as possible, staring into the dark cavern of the tank. A small reddish blob appears in my left eye, so I close my right eye to see it more clearly. The blob expands and grows darker at the center and scalloped along the edges. I am visualizing for the first time! And it appears that I have visualized a comic-strip sound-effect bubble that's missing its word. (POW! BOOM!)

To be honest, it's a little underwhelming.

I put my clothes back on and head to the front desk.

"How was it?" the desk lady asks.

"It was...interesting," I say. She waits expectantly, so I rack my brain for more things to say. "Very relaxing," I add.

As I leave my appointment, I am thrilled to see that the weather is back to being a nice spring day. I hop on a train to go visit a friend, and find myself questioning my experience. Did I *really* see something? Or did I just want to see something so badly, I convinced myself to get all excited about a phosphene.

I wonder if aphantasia comes with a tendency to doubt one's own sensory experiences — especially ones that can't be verified by someone else. Maybe visions are like Tinker Bell — they only exist if you believe.

My glimmer of a vision in the sensory deprivation tank reminds me of a strange evening I once had, in Gainesville, Florida.

The summer between my sophomore and junior years, I got a job as a research assistant at the University of Florida. My stated goal was to dip my toe into the waters of grad school, but my real motivation was to spend the summer with my boyfriend, Neil.

I'm not sure how I got off on such a wrong foot with Neil's roommates. It might have been when I told them that DJs are not real musicians. Or when I asked if they took party drugs to make their repetitive, thumping music more tolerable.

"Do you want to try some?" Neil's roommate Danielle replied.

"No thanks," I said, retreating to Neil's room.

Unfortunately, there was very little to do in that town besides drugs. So, my last night there, I just said yes. Danielle handed me half a gel tab of LSD and instructed me to let it dissolve on my tongue.

"It might take a while for it to take effect," she said.

People were beginning to arrive for a pre-party — a regular affair that usually culminated in me going to bed and everyone else going to a club. As the guests arrived, I found them all to be as vacuous as usual.

"You're majoring in business?" I asked a tightly wound guy. "How is that a major? What do you learn — how to make money?"

"Yes," the guy replied. "That's it exactly."

Defeated, I grabbed a bag of chips and retreated to Neil's room to read.

When I ran out of snacks, I returned to the party.

"Are you feeling anything yet?" Danielle asked.

"No, nothing," I said.

This was exactly what I'd thought would happen. If there's one thing you learn as a psych major, it's that people are extremely suggestible. If you give folks punch and tell them it's alcohol, they will start acting like drunk fools. *I knew it! All these expensive party drugs are sugar pills.*

"Are you sure?" Danielle asked.

"Yes," I said, rolling my eyes.

Danielle gave me another gel tab—a whole one this time—and she told me to check in with her in an hour.

The apartment was smoky, so I went outside and quickly became lost. Following the sound of thumping bass, I walked through an endless series of courtyards. So many grass lawns! So many brick buildings. So many cheap plastic porch chairs with the same broken leg. It took me a while before I realized I was walking in circles.

I would still be there today if a kindly moth hadn't taken pity on me. The big furry creature alighted on my shoulder, waggled his antenna, and then fluttered away—a clear indication that I should follow.

Mothman, my savior, my bat signal, a winged apparition in my hour of need, I thought, or possibly shouted.

I'm not sure how I made it back to the apartment, but I did—and I returned to my spot on the couch. How much, I wondered, had Danielle paid for these bogus drugs? *I hope they weren't expensive, because she will certainly ask me to pay her back.* My mind drifted to all the ways she had annoyed me that summer—general slovenliness, playing loud music, being hotter than me. But as I watched Danielle from across the room, I realized that the most irritating thing about her were her swirling paisley eyes.

I reached for a handful of chips, but the bowl dodged my hand. A woman sat down next to me, and I politely ignored the fact that she had snakes for hair. She asked me how I knew Neil and Danielle.

"Moooo," I said. I think I made myself clear.

Despite the mounting evidence, I persisted in believing that I was stone-cold sober. It took about two days before I remembered that I didn't normally converse with moths or spend forty-five minutes in the shower pondering the miracle of grout. Evidently, one and a half gel tabs is *a lot* of LSD.

I was right in thinking that expectations shape reality. I'd just forgotten that this rule also applies to me.

15

GULLIBLE'S TRAVAILS

Kent Cochrane was not your stereotypical mild-mannered Canadian. As a kid, he rode—and sometimes crashed—homemade dune buggies. As a young adult, he zoomed around Toronto on his motorcycle, staying out late and getting into the occasional bar fight. All this carousing came to a sudden end in 1981, when Cochrane skidded off an exit ramp and into psychology history.

The damage to Cochrane's brain was extensive. During his motorcycle accident, he had somehow smashed both sides of his head, obliterating both of his hippocampi—seahorse-shaped deep brain structures.

Despite this, Cochrane made an impressive recovery. He seemed almost normal, except for one glaring problem. He was unable to remember his rather colorful life, or make new memories.

He knew *how* to play cards, but he had no recollection of any particular card games. He knew *that* his brother had gotten married, but he didn't remember that he pranked his family by showing up with a big floofy perm. He knew *that* he'd gotten into a bar fight and broken an arm, but he had no personal memories of the event.

The famous psychologist Endel Tulving studied Cochrane and came to the conclusion that his peculiar brain injuries had robbed him of episodic memory—the ability to recall and relive experiences from your personal past. Cochrane's semantic memory—his memory for facts—remained intact.

With Cochrane's example, Tulving showed that these two types of memory rely on different brain structures. The hippocampi—the brain structures most damaged in Cochrane's accident—are clearly crucial to making and retrieving episodic memories, but they may not be necessary for remembering facts.

The hippocampi don't store your episodic memories. Rather, they take all the brain activation happening at a particular moment and bundle it together, serving as a kind of central hub to connect the various elements of an experience. The hippocampi encode this complex web of information, creating a unique "tag" for each event. This tag is then sent out to different areas of the brain where the details of the experience are stored. When it's time to retrieve a memory, the hippocampi reactivate these tags, pulling together all the distributed elements from across the brain to reconstruct the complete memory, much like piecing together a jigsaw puzzle. Thus, while they don't store the memories themselves, the hippocampi are crucial to our ability to remember and relive our experiences.

The ability to revisit our past and project ourselves into an imagined future is a mental feat that only humans can perform, Tulving wrote. By dint of his brain injuries, Cochrane had become stuck in an eternal present, unable to learn from his mistakes or plan for the future.

There's evidence to support the idea that other animals are stuck in the present. In one 1998 study, for instance, macaque monkeys and a chimpanzee didn't seem to care whether researchers gave them five bananas or ten. Either amount filled their bellies, and the primates apparently couldn't think ahead to a time when they might be hungry again.

In 2007, however, new research challenged this idea—so I called Tulving to get his take. We somehow ended up talking about his cat, Cashew.

"Animals are quite happy to do the Darwinian thing; you adapt to the environment as it exists," Tulving said. "My cat is very clever in many ways—she and her kind have survived tens of millions of years perfectly well. But they do not change anything about the world."

While cats may well be stuck in the present, scientists have found evidence of episodic memory in many other animals—scrub jays, apes, hummingbirds, and even lab rats. In the rat experiment, for instance, the animals were placed in a radial maze, where some arms contained treats, some of which had an artificial fruit flavor. After a one- or six-hour delay, the rats were returned to the maze. After the short delay, all the treats they'd eaten earlier were still gone. After the long delay, the fruit-flavored treats had magically replenished themselves.

The rats learned that the fruit-flavored treats only replenished after six hours, so they didn't bother looking for them after the one-hour delay. This proved that they could remember where (which arm of the maze), what (flavored or unflavored), and when (after one hour or six) an event had happened in their lives.[1] The researchers argued, then, that the animals had episodic memory.

While this was all well and good, Tulving said, the scientists had failed to prove that the animals had the subjective experience of mental time travel.

"But how can we know what rats are feeling subjectively?" I asked. "It's not like we can ask them."

"Tough luck. Certain problems in science are hard, but... they must be pursued," Tulving said.

After chatting with Tulving, I still felt very confused. What the heck is mental time travel? Is that, like, telling funny stories about your past? It must be, I decided.

In 2006, Susie McKinnon happened upon a biography of Endel Tulving and his work with Cochrane. In the book, Tulving posited that there might be people out in the world who were as incapable of mental time travel as Cashew—perfectly healthy and intelligent people who would have no idea that their lived experience was any different from the norm.

McKinnon immediately recognized herself in that description, and she emailed Tulving's colleague Brian Levine.

"I'm fifty-two years old, extremely stable, with a very satisfying life and well-developed sense of humor. Contacting you is a big (and, frankly, scary) step for me... I'll appreciate any guidance you may be able to give me."

Levine, a cognitive neuroscientist at the Rotman Research Institute and a professor at the University of Toronto, gets a lot of emails from people who think they have interesting neurological problems. McKinnon's stood out as both strange and plausible, so he invited her to come to Toronto for some brain scans.

You know how this story goes, so I'll fast-forward a little: McKinnon turned out to have the same lack of episodic memory as Kent Cochrane—but *without* the brain injury. She'd always been unable to relive particular moments from her past, and she assumed that when other people claimed to have detailed recollections, they were just making things up.

Along with McKinnon, Levine invited two other folks who'd reached out to him to Toronto for testing. They all claimed to lack episodic memory—a phenomenon that Levine ended up calling severely deficient autobiographical memory, or SDAM.[2] (Sadly, "developmental amnesiac" was taken.*)

The participants took a battery of tests—to measure motor skills, executive function, general intelligence, working memory, visual memory, and more—and the compatriots did as well as control participants in all cases, except one that involved drawing a complex figure from memory.

Another major difference between SDAMers and control participants showed up in a test of autobiographical memory. The three SDAMers were surprisingly good at remembering details from recent life events, but the number of details fell off quickly for things that happened more than a few months in the past. For these older memories, control participants recalled more than twice as many specific details about the events.

While the SDAMers' brains appeared largely normal, Levine noticed one subtle difference. Their left hippocampi tended to be larger than their right hippocampi. Neurotypical brains generally show the opposite asymmetry.

The right hippocampus is more involved in consolidating the visual aspects of memory, while the left hippocampus links more to language centers. This pattern of asymmetry fits with the fact that neurotypical people mostly store memories as images, while

---

* Developmental amnesiacs are people whose hippocampi were damaged at a very young age. They have significant learning and memory problems as a result.

SDAMers use language or perhaps some other, more abstracted process.

About one-third of people with aphantasia also have SDAM. I think I'm in that third, so I connect with Levine via a video call.

"How did you come up with the name for SDAM?" I ask.

A pained expression crosses Levine's face.

"It's a terrible name. I regret calling it that," he says. "It's impossible to remember, and it focuses on the deficiency, when actually it seems like there's a whole profile of strengths and weaknesses."

Levine chose SDAM so that it would bookend the opposite memory condition, HSAM—highly superior autobiographical memory. People with HSAM remember almost every day of their lives—name a date at random, and they will tell you what day of the week it was, what the weather was like, and what they had for breakfast. It's an incredible gift, but it has a clear downside. Overwhelmed with irrelevant information, people with HSAM often get mired in details and have trouble moving on from the past. SDAMers, in contrast, excel at big-picture thinking and are almost incapable of holding a grudge.

These two groups have perfectly healthy brains, Levine emphasizes. They simply inhabit the extreme ends of a mnemonic spectrum. HSAM folks put all their chips into episodic memory, while SDAMers put theirs on the semantic side. Neurotypical people strike more of a balance between these two kinds of memory.

Episodic memory captures the details of a particular event, while semantic memory picks out commonalities, Levine says. Imagine an SDAMer and an HSAMer are on a hike together, and they see someone get bitten by a dog. The HSAMer will remember the breed of the dog, the time of day, the weather, and the smell of pine trees. The SDAMer will forget all these things while pulling out general principles—that dogs attack when they feel threatened, for instance, or the characteristic behavior of a scared dog.

This is why people with SDAM experience difficulty memorizing information by rote. Compared to people closer to the HSAM side of the spectrum, those with SDAM brains refuse to hold on to information unless we can fit it into a bigger picture. Like Craig Venter said, there are no shortcuts for us. To remember something, we have to understand it.

When we spoke on the phone, Venter said he'd never heard of SDAM, but I think he has it. I suspect that many aphants do. The relationship between the two phenomena is unclear, Levine says, but it does seem that the ability to vividly visualize events from your past is crucial to capturing all those details.* It's possible that SDAM represents the extreme end of aphantasia—people who barely encode visual information to begin with. Aphants without SDAM, in comparison, might be saving visual information but be unable to retrieve and replay it later. While they can't consciously visualize, they may be able to unconsciously tap into some of those remembered details.

It's also possible that people with aphantasia without SDAM simply don't realize how good other people are at recalling the details of old events—though if you tested them, they would still fall short on internal details, Levine says.

All of this is really clicking with me. I've always felt like other people are hung up on irrelevant details. I call all coffee shops Starbucks and all painkillers Tylenol. One time, when my pharmacy started giving me the wrong birth control pills, I failed to notice for months. I never remember the names of characters in books or movies (who cares? they're made up!), but I excel

---

* In sighted people. No one has studied these phenomena in the vision impaired yet.

at pulling out overarching themes. The plot for *Memento* eluded me, but I did notice that the black-and-white scenes took an objective perspective, while the color scenes appeared to be tinted by the main character's feelings and beliefs.

I've never met Levine before, and from his descriptions of SDAM, I feel like he knows me. This emboldens me to ask a painfully embarrassing question.

Do you recall Neil, the boyfriend I trained like a lab rat? I later found out that the reason he'd been missing our calls is that he was cheating on me. He kept this up for nearly four years of college, without me having a clue. Whenever I visited, a "friend" of his would park in his driveway and sob. Neil vaguely explained that she was jealous he was spending so much time with me. "Your friend is really intense," I observed.

When I finally figured it out, my stepmother, Lynn, comforted me by saying that now that I knew the signs, I wouldn't be so oblivious in the future. She was wrong. My very next boyfriend, "Ian," cheated on me for eight months—and I was completely caught off guard. Again.

Neither of these guys were criminal masterminds. They barely had to cover their tracks. Ian, who lived with me, would sometimes disappear for days and return with wild stories that invariably involved losing his phone. Since I, too, tended to have unlikely adventures and regularly misplaced my phone, this seemed plausible to me. Plus, there was enough time between these incidents for me to forget about the previous one by the time the next one happened. I had a vague sense of unease, but since I had no concrete memories to back it up, I dismissed my feelings as irrational.*

---

* Interestingly, Venter emphasized the importance of trusting your gut. This may not be good advice for everyone, but I bet it is for people with SDAM.

"Do you think that maybe I'm not very good at learning from my past experiences? Because I can't remember them? Or I remember, but my memories are too detached from my emotions?" I say.

Levine comes up with a very polite way of saying, *No, Sadie, you can't blame your naivety on SDAM.*

"That might be something that arises in a more complicated sort of scenario," Levine says. "Because we might predict that even if you don't remember the episode, you should remember the factual component—'to avoid people who do X...'"

"Right, right. That makes sense..." I reply. My sinuses are burning. I hastily thank Levine and hang up before the waterworks start.

I'm not upset because of any specific thing. I'm upset over *everything*. All the emotions from every time in my life I've ever felt betrayed have decided to bubble up at the same time.

Most people would say that my current meltdown is happening because I didn't deal with my feelings the first time around, that I shoved them down. And that's true to some extent. But while neurotypical people have to work to suppress negative feelings, mine evaporate on their own. Sort of.

When I found out about Ian's secret life, I was a mess. I ran to Sybil's apartment, threw myself on her couch, and sobbed for hours. Then I returned home and instituted my patented breakup procedure.

I deleted Ian's phone number and banished his photos to the back of my closet. I blocked him on social media and began avoiding places where he might be. Without any reminders, our entire relationship was a Polaroid photo left out in the sun. I forgot the whole thing, good times and bad, in a matter of months.

Remember how visualizers in Pearson's study got stressed out while reading the story of a shark attack, while the aphants did not? That finding suggests mental imagery evokes strong emotions that words alone can't produce. Since there's no imagery attached to my memories, I suspect the related emotions easily become detached. If I wanted to keep uncomfortable emotions alive and accessible, I would have to wallow — on purpose. I'd have to get out all the photos and mementos and have a big ugly sob session. This kind of masochism runs completely against my upbringing and my nature.

Later that day, I realize that letting my emotions evaporate on their own is effectively the same as when neurotypical people repress their feelings. The feelings are still around, but instead of helping me learn specific lessons, they are just hanging out, waiting for an opportunity to bubble up and ruin my day.

The next time something awful happens, I might try wallowing, just a little bit.

After I pull myself together, I email Levine and ask if he needs any more (likely) SDAM brains for his research, and he says yes! Toronto, here I come.

Baycrest, home of the Rotman Research Institute, is a Jewish retirement community that sprouted a hospital and then a research center. I'm in an indoor courtyard sitting underneath a tree when Levine appears. With his curly blond hair and impish smile, he reminds me of Gene Wilder.

We head upstairs and settle into a ludicrously tiny room for a slew of tests — many of which I vaguely recall from the times when a teacher decided I was gifted or in need of special education and sent me to the school counselor for testing. There are strings of numbers for me to repeat and lists of words to remember. We also play a card game with secret rules that I'm supposed to figure out. As always, my scores are boringly average.

I'm doing fine until we get to the Rey-Osterrieth Complex Figure test, which I later find out is the same test the SDAMers in Levine's 2015 study flunked. The way it works is, you spend a few minutes looking at a picture that consists of geometric shapes glommed together with a few little flourishes—a box filled with dots here, a stray line there. I stare at it for a few minutes and then try to draw it from memory. All I remember is the basic outline. I thought I did pretty well until I saw the drawings of people who scored "average" on the test. The amount of detail most people can remember is frankly astounding.* I'm feeling increasingly convinced that other people really are able to store pictures in their minds and retrieve them later.

After the tests, we have lunch on Baycrest's roof with some members of Levine's lab. They have a lot of questions for me, and I get it. If I have SDAM—and I'm pretty sure I do—that means I have very little by way of episodic memory. That puts me in the company of the protagonists from *Memento*, *Finding Dory*, and *50 First Dates*—characters who can't recall what they're doing from moment to moment or day to day.

In real life, people who lose their episodic memory to brain damage are not constantly disoriented. They can remember what they're doing for as long as they're doing it. Cochrane, for instance, could play long games of chess. But when he wasn't absorbed in a particular activity, he'd forget anything that happened more than fifteen minutes ago.

People with SDAM don't have that problem—which suggests that we have some special workaround that we developed as children.

---

* The test makers won't let me show you for fear that people will study the figure and ace the test.

Indeed, Levine has found that SDAMers' *recent* memories contain the normal amount of what he calls "internal details"—information specific to an event, such as the setting, what people were wearing, or what emotions we felt at the time. After a week or so, however, SDAMers start rapidly losing that information and—since sensory details trigger reminiscing—we eventually lose touch with the memory entirely.

Losing one's memories—to Alzheimer's disease, for instance—is a horror that most people don't care to contemplate. Memory is the foundation of our personality and our identity. Memories allow us to reflect on the past and gain wisdom from mistakes. There's also evidence that without memory, we are unable to imagine the future. When Tulving asked Cochrane what he might be doing a year in the future, he drew a blank. "It's like swimming in the middle of a lake. There's nothing there to hold you up or do anything with," Cochrane replied.[3]

In comparison, people with SDAM are healthy and functional. Indeed, we are so good at compensating for our lack of episodic memory, we don't even realize that our brains are radically different from the norm. One way I suspect I do this is to write myself a lot of notes. Whenever I finish writing for the day, I leave myself a message about what I accomplished and what I need to do tomorrow.

Modern technology is also extremely helpful. I record all appointments in my Google Calendar because there is literally no chance that I will remember them otherwise. I take a lot of photos, and in the pre-digital era, I got them developed and made myself huge photo albums, with notes about the people and events they captured.

But I'm beginning to suspect that my central compensation also happens to be my job: storytelling.

Stories, told in words, are what I use in lieu of episodic memory. Whenever anything interesting or weird happens to me, I tell the first person I see. Then, as soon as I'm near a computer, I'll record the incident in the form of a long, entertaining email to a friend or jot it down in my journal. Sometimes these stories make their way into magazines or newspapers. So while I can't recall a single moment from my wedding, I can always read about it in the archives of the *Washington Post*.

According to psychology professor Dan McAdams, we are all constantly making sense of our lives by putting otherwise confusing or random events into the form of a story. People who tell coherent stories—stories with foreshadowing and clear chains of cause and effect—tend to score higher on a variety of measures of psychological well-being, including resilience and life satisfaction.[4]

Researchers have also found that people who tell stories with happy endings tend to be happier—but unless you also include moments of struggle, you will miss out on an opportunity for personal growth. In one study, a team of psychologists at Southern Methodist University asked parents to tell the story of how they learned that their kids had Down syndrome. The parents who told stories with happy endings actually grew happier over the course of two years. So, essentially, they told the story and then lived their way into it.[5]

Only some of these parents, however, also showed gains in ego development—a measure of the complexity and sophistication with which a person views the world. What kind of stories did these parents tell? They included moments of struggle and conflict. In essence, they didn't rush right to the happy ending.

Like Taylor Swift, I find myself growing older without getting wiser. I think the kinds of stories I tell may be to blame. I like comedies—stories where the main character succeeds despite, or perhaps even *because of,* her inherent flaws. Nothing wrong with that, but I tend to gloss over very real moments of struggle and confusion. This is perhaps why, for decades, I failed to see the throughline of my many zany adventures. My oversimplified stories blinkered me to a major truth about myself: that I am, emphatically, not neurotypical.

16

COMIC INCOMPETENCE IS THE BRAND

Enterprise, Hertz, Avis, Budget, National, Dollar, Thrifty... I'm at the Tampa International Airport, picking up my first-ever rental car. That is, if I can find the company I reserved the car with. I've reached the end of the last counter, and ACE is nowhere in sight.

I ask a Budget employee for help, and she points me down a long, dark hallway, at the end of which is an empty desk.

"Hello?" I call into the void.

A man in a dress shirt pops up from behind the desk.

"Heya!" he says. "Is this your first time in Tampa?"

"No," I reply. "I grew up here."

He did too, and as it turns out, we both graduated the same year from rival high schools.

"What are you in town for? Visiting your folks?"

"Yes, but I'm also going to a vision science conference. I'm lucky — it's at a beach resort that's just ten minutes from my dad's house."

"Well, I won't ask you if you know your way around," he says with a chuckle.

I chuckle too, but I'm lying. As a kid, I never paid a lick of attention to the streets or roads. Why bother? Someone else was always schlepping me around.

There were a few major downsides to being an eternal passenger. I was always at the mercy of other people's schedules and music preferences. And if I didn't pay attention (and I didn't), I would be dragged on unauthorized side trips to Home Depot (if my dad was driving) or Best Buy (if Neil was behind the wheel).

It's going to be fun to tool around in my hometown, captain of my own ship, a vessel that only detours for iced tea. I'm a little anxious, though. I've been driving for more than a year now, but most of my experience is on deserted country roads. Tampa is known for its complicated thoroughfares and incompetent drivers. But I have a plan.

As soon as I'm out of the rental-car employee's sightline, I stop, pull two STUDENT DRIVER magnets from my backpack, and slap them on the car.

A family of three stops and stares.

"Aren't you a little old to be a student driver?" asks the mom.

"I am a lifelong learner," I reply.

On the way to my grandma's apartment, the GPS lady reels off names of streets that I've heard many times before but never attached much meaning to: Kennedy, Bay to Bay, Dale Mabry.

Personal landmarks also keep appearing unexpectedly. My high school! My favorite bagel place! Why didn't anyone tell me they were so close together?

COMIC INCOMPETENCE IS THE BRAND

My grandma's senior-living apartment complex is nearly in sight when an aggressive toll road sucks me in and deposits me on the wrong side of town. I find myself in what appears to be a zombie shipyard. Great job, Tampa city planners. I'm sure many people appreciate the convenience of the downtown-to-industrial-hellscape expressway.

I have no idea how to get back downtown, but my smartphone does. What did people do before GPS-based driving directions? Maybe it's a good thing I didn't learn to drive as a teen.

When I finally park at my grandma's place, I get out of my car and see that I am not at all between the lines. Should I try again? Nah, I decide. People will see my STUDENT DRIVER bumper stickers and understand.

As I walk through my grandma's building, it occurs to me that my "student driver" label comes with some unintended consequences. By asking other people to hold me to a lower standard, I give myself permission to suck. Is this true of the other labels I've been collecting?

I find my grandma sitting on her couch, reading a book about Frank Sinatra.

"It's very interesting—he was born in a time before everyone had radios at home," she says. "If you wanted to hear music, you had to go see a live show."

"How did he get into show business?" I ask.

My grandma thinks for a minute and comes up empty.

"I'm very old," she says, looking embarrassed. "My memory is not so good."

"What are you talking about?" I say. "Your memory is amazing. What was your address when you were a little girl?"

My grandma rattles off some New Jersey street address without a moment's hesitation.

"See? That's amazing. I can't do that!"

My grandma's fading memory has many of my family members worried. I'm not among them—I think her memory is *amazing*, especially for someone who is ninety-four. If you pull out old photos, she can name every person in them—including distant cousins she hasn't seen for nearly a century. She reads the newspaper every day and is a lot more on top of current events than I am. Okay, so she can't keep track of her hearing aids—but that's not her fault. Those things are tiny!

I wonder if my neurodivergent brain allows me to see ability where other people see impairment.

My grandma interrupts my reverie to tell me about this new book she's reading.

"The interesting thing about Frank Sinatra," she says, "is that he was born before everyone had radios at home."

"That *is* interesting," I say. "How did he get into show business?"

(It's possible that I am delusional.)

I walk into the Vision Sciences Society (VSS) conference practically vibrating with anxiety. Almost everyone I've roped into my nerdy midlife crisis is here, and there's a good chance I won't be able to recognize any of them. Plus, I'm beginning to feel sheepish about this whole endeavor. Writing an autobiography is self-indulgent, it's navel-gazey, it's all the things I hate. My life is rather ordinary. Is my brain really all that special?

When I asked Levine's postdoc H. Moriah Sokolowski this question, she sent me an interesting paper.[1] It detailed a study in which scientists at the University of Cambridge collected an impressive heap of data on 347 students, ages five to nineteen, who had been diagnosed with neurodevelopmental disorders, such as dyslexia, ADHD, and autism. Another 142 students from surrounding schools were used as controls. These kids took a lot of tests,* and many of them also hopped into an MRI for structural brain scans—using a new technique that traces connections between different parts of the brain.

The scientists then fed all this data into a machine-learning algorithm and asked it to find naturally occurring groupings of kids with similar cognitive profiles and kids with similar brain structures. The program also looked for relationships among all the variables.

Once it was done processing all this information, the program was able to find only two groups of kids: neurotypical and neurodivergent. Just two! What it *didn't* find were groups that corresponded to *any* of the labels we know and love: dyslexia, autism, ADHD, etc. These labels, the computer said, are meaningless.

In terms of brain structure, there was a noticeable difference between these two groups. Neurotypical kids have brains that employ a hub-and-spoke network. This is a highly efficient system in which a few highly connected hubs link out to more sparsely connected areas. If a neurotypical brain were an airline, you'd never have to make more than one connecting flight.

The neurodivergent kids, on the other hand, have more of a daisy-chain network, a type of network that results in a few ultra-efficient connections and others that are quite inefficient. If neurodivergent brains were airlines, there would be a few random express flights—from Miami to San Francisco, for instance—but to get to San Diego, you'd have to make several stops on the way.

---

* These included measures of working memory, fluid and crystallized reasoning, phonological processing, verbal and visuospatial memory, and reading and math fluency.

This is perhaps why we neurodivergent folks often have uneven and perplexing cognitive profiles. I have no trouble pulling out multiple vocal lines in a piece of music. But two people talking at once? They may as well be speaking French.

The idea that we neurodivergent people are really one big quirky family makes a lot of sense to me. Under the current system, most of us qualify for multiple labels. Upwards of 40 percent of kids with ADHD also have dyslexia.[2] Around 50 to 70 percent of kids with autism also have ADHD.[3] Around 37 percent of autistic kids are also faceblind.[4] I'm not just faceblind, stereoblind, aphantasic, and SDAMish. I also have topographical agnosia, left-right confusion, and (I suspect) a touch of auditory processing dysfunction. The psychiatry bible known as the *DSM* (*Diagnostic and Statistical Manual of Mental Disorders*), however, would tell you that I'm entirely neurotypical.

According to Sokolowski, there's mounting evidence that our current system of labeling neurodevelopmental (and mental) disorders is garbage. Indeed, the very idea of sorting brains into buckets may be flawed. There are probably a zillion ways for a face-recognition system to fail, and two completely different failures might result in the exact same apparent deficit. The brain has 86 billion neurons with an average of 2,000 connections apiece—and the whole system is moderated by a poorly understood chemical bath. We humans simply haven't devised a way to understand systems with this kind of complexity. Maybe machine learning and artificial intelligence can help us figure it out.

And yet, I love my neuro-buddies! I have gotten so much out of meeting my fellow prosopagnosiacs, aphants, stereoblind folks, and SDAMers. When you feel like an alien trying to fit in among humans, finding other aliens with similar difficulties makes you feel so much less alone. We commiserate, tell stories, and trade tips! Together, we try to figure out how to hack it in a world that was not made for brains like ours. Or, even better, we work on reshaping the world so that it's hospitable to all sorts of unusual brains. There are so many needlessly constraining social norms. Why shouldn't we talk to ourselves in public? Why can't I wear sensory-friendly clothing to work? Why do we expect people—children and adults—to sit in chairs for eight hours a day? A more flexible, less judgmental world benefits everyone.

Perhaps this is why the burgeoning neurodiversity movement seems to be developing a big-tent philosophy. You don't have to have an "official" diagnosis to call yourself neurodivergent—doctors are expensive, and not everyone is as good at getting into studies as I am. You may feel that your experience is perfectly encapsulated by a *DSM* description, or maybe your cognitive profile defies categorization. Even if you find labels useful, your version of, say, autism might be wildly different from someone else's. Brains are complicated, the world is equally complex, and everything we think we know is provisional.

Ever since learning that I am faceblind, I have been trying to stop pretending that I know who people are—and interestingly, this has made me more sensitive to other people's BS. In DC, for instance, you can't throw a rock without hitting someone who is making confident statements about a topic they obviously know nothing about.* "Agnosia," in its original Greek form, means not knowing, and I think the world would be better off if we all could be a little more pro-agnosia. Admitting what you don't know is, after all, the first step toward figuring it out.

---

* This turns out to be easy to spot. Just look for people making definitive statements. Real experts refuse to be pinned down on anything—and this can be quite frustrating to science writers like me.

DO I KNOW YOU?

* * *

While I'm increasingly concerned about the validity of labels, they are certainly useful — especially if you don't want to be constantly explaining yourself. So, for VSS, I have made myself some stickers. They feature a befuddled-looking golden retriever and the text I'M FACEBLIND. I'M JUST PRETENDING TO KNOW WHO YOU ARE.

My sticker is an immediate hit. Before I even go into my first session, several people ask me for one. "Are you faceblind too?" I reply. I get a variety of answers, from "Definitely" to "Nah" to "Maybe" — but they all want stickers. I'm not sure how I feel about this, but if it results in general faceblindness awareness, or even just everyone introducing themselves to everyone else, that works for me.

The first session I attend is on face perception, and the room is packed. I sit down in the third row and immediately spot a celebrity: Brad Duchaine, the guy who basically discovered developmental prosopagnosia.

Duchaine's grad student, Antônio Mello, gives the first presentation, and it is riveting. Mello has been working with a fifty-eight-year-old man, "VS." A few months after suffering carbon monoxide poisoning, VS developed full-face prosopometamorphopsia (PMO), a condition that causes faces to look distorted, as if they are swelling or melting. (You may recall hemi-PMO from chapter 2, a condition that causes the left or right side of faces to appear distorted.)

"Everyone looks like demons. Basically, the whole face is kind of stretched backwards. The eyes are more narrow, the ears are distended, and there are lines down the cheeks and across the forehead," VS told Mello.

Interestingly, this distortion only appears on faces VS sees in 3D — in real life. Faces on screens or paper continue to look normal.

Mello realized that VS's condition meant that he could look at a photo of someone's face on a computer and adjust that image until it exactly matches what he sees in real life.

"It's not often that you can ask someone to provide a photorealistic depiction of their subjective experience," he says.

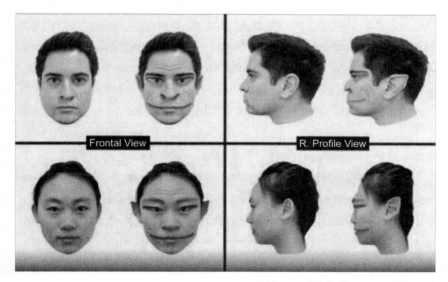

(Antônio Mello, Dartmouth College)

Mello and his colleague Elizabeth Li took turns sitting in front of VS while he adjusted the computer-screen images of them until they matched what he saw. The fact that VS's pattern of distortions is extremely consistent across different people and different viewpoints supports the theory that normal face perception involves fitting every face into a standardized 3D template. PMO seems to result from a glitch in that step of the face-recognition process.

VS finds these distortions disturbing, but he's discovered a way to make them disappear. When he wears green-tinted glasses, faces look almost normal. Red-tinted glasses, in contrast, exaggerate the effect.

Mello and his colleague have found three other PMO patients who also report this color effect—and he hopes their work can help people with this rare disorder and advance our understanding of face perception.

Research on people's subjective experiences is unusual, so I'm surprised to see another presentation on the topic a little later in the session. According to preliminary research by University of Haifa grad student Marissa Hartston, on average, autistic people who also have prosopagnosia have trouble encoding faces, but their ability to retrieve faces from memory seems largely unimpaired.[5]

When Duchaine asks the audience if we have any questions, my hand shoots up. "Do faces look different to people with autism—like, subjectively?" I ask.

A look of consternation crosses Duchaine's face. What, I wonder, did I do wrong? Are we not supposed to use the *s* word?

"That's a really interesting question," Hartston says. "I don't have a definite answer, but what I can say is that if I imagine myself landing in the middle of Beijing, I think I'm going to have a hard time differentiating between faces. And what has become clear to me is that these individuals with autism—that's how they spend every day of their lives...trying to differentiate all faces the same way we differentiate other-race faces."

When I hear the next question, and the person's voice is obviously amplified, I figure out my mistake. Behind me is a microphone with a queue of folks standing patiently in line. I accidentally cut in front of everyone by shouting my question from the audience.

After the talk, Duchaine pulls me aside. "Next time you have a question, there's actually a microphone..." he starts to say.

"I know," I say, cringing. "I mean, I figured it out, a little too late."

COMIC INCOMPETENCE IS THE BRAND

From Duchaine's generic, polite demeanor, it's clear that he doesn't recognize me. Ha! I guess it happens to everyone.

Later that day, I attend the opening night dinner and sit with a table of grad students. They are welcoming but not very chatty, so I wander to the beach. It's a gorgeous evening—the setting sun has turned the sky into a swirl of blue and tangerine. Just beyond the waves, the water is glassy flat. A flock of birds passes by, flying low, dragging half of their candy-corn bills in the water.

"Skimmers!" I exclaim, to no one.

To get a better view of the black skimmers, I wade into the ocean. A shimmering school of mullet slaloms through my ankles. Tiny terns hover above my head and dive-bomb all around me. My heart swells with joy at being immersed in all this natural beauty, and this feeling is accompanied by a pang of regret. I used to live near this beach, but my inability to drive kept me stuck at home.

A tourist wades over to me. "How do you know there aren't sharks in here?" she asks.

"Oh, there are definitely sharks," I say.

Shark attacks are rare, but if you were trying to get bitten by a shark, you'd do exactly this: stand in a jetty at sunset, in thigh deep water, amidst schools of fish.

On cue, a lemon shark cruises by. He's just a little guy, but we both take it as a sign to get out of the water.

As I walk back to the conference center, something disgusting catches my eye. Three fish heads, probably discarded by a fisherman, have washed up on the beach. I don't want to look, but I'm drawn in by their gross, swollen eyes—inflated by decomposition gas, I suppose.

My inner DJ is not going to let this moment pass without commentary. Do you know the fish head song? It goes, "Fish heads, fish heads, roly-poly fish heads..." It's quite catchy, and it's going to play in my head, intermittently, for the rest of the week.

This is par for the course, for me—but what's weird is that this song is sometimes accompanied by a visual. It's very faint, and it disappears if I scrutinize it, but it's clearly three fish heads, with blind bloated eyes.

I guess this is what visualizing is like? It's less fun than I'd hoped.

Throughout the conference, I keep checking my email, hoping for a message from Wilma Bainbridge or Brian Levine. Both professors are analyzing my brain scans, and any day now they'll let me know if I officially have aphantasia and/or SDAM.

Weirdly, I'm hoping it's a yes on aphantasia and a no on SDAM. This doesn't make a lot of sense, since there's so much overlap. The two labels may capture a difference in severity, or perhaps they are simply different angles on a common cognitive profile. In any case, it feels like having aphantasia is kind of cool, while having SDAM is a tragedy.

I may have picked up these attitudes from online message boards. The aphants on Reddit are curious about what visualizing is like, but they are generally happy with their brains. Whenever someone posts about "curing" aphantasia, it attracts a lot of criticism, or is quickly voted down. The general consensus is that aphantasia isn't a disorder; it's just the extreme end of the imagination continuum, with the hyper-visualizers on the other end. Sure, we're unusual, but that makes us valuable.

Meanwhile, the SDAM folks are markedly less sanguine. On their subreddit, I see a post comparing the condition to Harry Potter's Dementors, which suck happy memories and good feelings from anyone who gets too close. Another SDAMer writes that she feels "stuck in the present and disconnected from myself." A third wonders if he is cold-hearted because he gets over break-ups so quickly, asking, "Is it possible to 'love' in the traditional sense with SDAM?"

People's responses to learning they have SDAM seem, to me, to encapsulate the old-school, deficit-focused view of brain differences. The aphantasia response, however, reflects a more enlightened perspective, one that approaches brain differences with curiosity instead of judgment. Yes, being different can make life hard at times, but instead of trying to change neurodivergent brains so they can fit into society, maybe we should change society to accommodate neurodivergent brains.

These ideas dominate my thoughts throughout the conference, and I find myself buttonholing scientists to warn them about the power of labels. If you discover a neurodivergence that resonates with people, especially one that gives a name to an inchoate feeling of being different, it might catch on. Before you know it, a whole community could spring up around a word you made up. So, please choose wisely.

When I was a kid, my dad and I used to love crashing hotel pools. The trick is to act like you belong. Specifically, you must travel light. Do not, under any circumstances, bring your own towel. If the towel stand is manned by someone asking for room numbers, just nab an abandoned towel off of a chair.

"No one really looks at anyone," my dad said, "unless you give them a reason to."

Maybe *this* is where I get my belief that I'm basically invisible. I have a lot of experience sneaking into places I don't belong—and getting away with it.

My dad must be nostalgic for our hotel-crashing days,

because he keeps dropping hints about tagging along with me to the Vision Sciences conference. "Really? You want to listen to lectures about eyeballs?" I ask. "I want to meet your friends," my dad says. I eventually bring him to "demonstration night" at the conference, and I don't tell him that family and friends are explicitly invited.

After a barbecue dinner on the beach with DeGutis and company, my dad and I find a room ringed with conference tables. At each table stands a vision scientist demonstrating a teaching tool or an optical illusion. We join a crowd of people who are watching projections of rings that appear to be moving when they are not — and I have to say, the effect is rather subtle.

"Did *you* see anything?" I ask my dad.

"Not really," he says.

We only see a few more demonstrations before my dad's knee starts acting up. "I'm going to find a place to sit down," he says. "Okay, I'm going to the bathroom," I reply.

I'm heading in that direction when a professor buttonholes *me*. She's late to set up her table, and she seems very vexed. When I smile at her, she hands me three pieces of paper featuring tiny optical illusions.

"These images, they are too small," she says, breathless. "Can you make them bigger?"

"I don't actually work here," I say.

"Can you do it?" she presses.

"Okay," I say. "I'll try."

As I look around the conference center for a copy machine, it occurs to me that I'm uncommonly go-with-the-flow. Is this trait, I wonder, a consequence of growing up faceblind? It's possible. Maybe I've gotten used to people I don't recognize telling me what to do.

COMIC INCOMPETENCE IS THE BRAND

No, this doesn't ring true. I'm not known for being compliant and following directions. What I am known for, however, is diving into ridiculous situations. Even if they don't turn out to be fun, they make good stories.

Learning about my various neurodevelopmental conditions has helped me see my life more clearly, but they don't explain everything. My lifelong mission — to have interesting experiences and turn them into stories — isn't a prosopagnosia thing, a stereoblindness thing, an aphantasia thing, or even a Dingfelder thing. It's my own invention.

Behind an unattended desk, I find a room with computers and a copy machine. It could be a business center for public use, but it could also be someone's office. Hard to tell.

"Hello?" I say as I step inside. There's no answer, so I blow up the optical illusions, copying them onto huge pieces of paper.

On my way out, I notice a credit card reader on the machine — one that I seem to have accidentally circumvented. Oops.

Finding loopholes is another specialty of mine — and I think this may be a neurodiversity thing. If you live in a world that was designed by people with abilities that are wildly different from your own, you have to be scrappy and inventive, just to get by.

I return to the conference room and triumphantly hand the professor her now poster-sized illusions.

"This is much better," she says, sounding deeply relieved.

Then she returns to her conversation with another professor. Where's my thanks? My cookie? My parade?

I find my dad sitting at an empty table — where some demonstrator has failed to show up. There's an unused power strip, an extension cord, and a placard that says, LET'S TEST THE POLARIZED CONTRAST THRESHOLD!

"I think I just experienced what it's like to be a grad student," I say.

A pair of students walks up to us and asks to see our demonstration. My dad starts pushing around the power strip. "Do you see it yet?" he says.

"Ignore him," I say. "We're just sitting here."

As someone who is very gullible, I feel a duty to protect innocent academics from my dad's shenanigans. But that feeling quickly dissolves when a new group walks up to the table, and peers over my shoulder at an empty corkboard behind me.

"Do you see the fixation cross?" I say, while pointing to a random spot.

The students shake their heads.

"Sadie, stop it," my dad stage-whispers.

We decide that we'd better leave before we get ourselves kicked out. We're on our way out of the convention center when a tall woman spots me and rushes over. According to her name tag, she's Emma Megla, from Wilma Bainbridge's lab.

"It's so good to see you," I say. I'm only able to manage about five minutes of small talk before asking if they have my results yet.

"Not yet," Megla says. "But I will have them soon."

In academia, "soon" can mean anywhere from tomorrow to never, so we make a date to go over my results on a video call.

"Thanks so much," I say. "I can't wait!"

A month after the conference, I get on a Zoom with Megla and Bainbridge. Megla has prepared a slideshow, and it's all about me! We look at images of my brain next to the brain of a matched control, to see how our brains activated when we looked at pictures and then tried to visualize what we had just seen.

As expected, both of our brains activated normally when we were looking at pictures—faces activated both of our FFAs, and places activated both of our OPAs (occipital place areas).

The difference happened when we visualized. When I tried to imagine things, my brain appeared to do nothing. When the control participant visualized, the same visual processing areas lit up—though not as strongly as before.

It's official. I'm an aphant. Time to order my T-shirt. Do we have a flag? If so, it should be blank.

"This is what I was most interested in, because it's a little bit of objective data that I'm correct in saying that I'm not visualizing," I say.

"This is really exciting for us too," says Wilma. "This gives us some objective evidence that this is a true thing, and not just that people don't understand their own imagery, or are overestimating other people's imagery."

A few slides later, Megla shows me a graph of the functional connections between my medial temporal lobe (the area that includes my hippocampus) and the rest of my brain. Compared to the control participant, my connections are quite sparse.

"In other areas of your brain, like the brain stem, your connections are normal," says Megla. "But we see fewer connections between your medial temporal lobe and areas like your prefrontal and temporal lobes."

This fits perfectly with the White House press secretary analogy I mentioned before. I often get the sense that my conscious mind is just making up explanations for decisions it had nothing to do with.

"Do I have any extra connections anywhere?"

Wilma says that we might expect to see more connections between my medial temporal lobe and the semantic, knowledge-storage areas of my brain—but those are hard to capture because they are blocked by the ear canal.

Not long after our chat, I hear from Brian Levine, and his results dovetail nicely with Bainbridge's findings. As you may recall, neurotypical people generally have a bigger right hippocampus as compared to their left. I show the opposite asymmetry: my left hippocampus is larger than my right. Additionally, my fornix (the bundle of fibers that connects my hippocampi to the rest of my brain) is a bit threadbare. Both of these structural differences are typical of folks with SDAM.

This feels like confirmation that I really am neurodivergent. My brain isn't just acting weird; it *is* weird. It's literally wired differently than neurotypical brains.

My feelings of validation suddenly give way to an embarrassing realization: I am an amnesiac memoirist! This is even worse than being a faceblind reporter.

I feel panic rising in my throat. What does this mean for the book I'm writing? Will the whole project be canceled? Am I going to have to pay back my advance? Maybe I should just leave this part out...

"Are my memories less accurate than other people's?" I ask Levine. "Would you believe a memoir written by someone with SDAM?"

"You're not going to remember as many details, but what you do remember is going to be as accurate as anyone else—maybe even more accurate," he says.

I sure hope he's not just being nice. Now it's Levine who looks anxious, like he's afraid I'm going to try to talk about my personal life again.

"You probably have compensations..." he adds.

COMIC INCOMPETENCE IS THE BRAND

"I do!" I say. I snap pictures, I record interviews, and I take loads of notes.

Indeed, I have always been a compulsive chronicler of my own life — a journalist long before I knew it was a career.

This leads me to the final, and perhaps biggest, revelation of my nerdy midlife crisis. I don't *have* prosopagnosia, stereoblindness, aphantasia, and SDAM. I *am* all of these things. They are the grains of sand in my oyster — irritants around which I have built the pearls of Sadie-ness. Prosopagnosia gifted me with intense friendliness and comfort with ambiguity. Stereoblindness gave me the perspective of an eternal outsider. SDAM and aphantasia forced me to become a storyteller and a writer, simply to remember important moments that I knew I'd otherwise forget.

I've been grappling with who to dedicate this book to, and I finally figured it out. I'd like to thank my weird brain — one that, for forty years, fooled me into thinking that it was entirely neurotypical.

APPENDIX

Practical Advice

While working on this project, I picked up tips from the professionals as well as fellow faceblind folks—practical advice that I didn't want to shoehorn into my personal story. Here it is; I hope you find it useful.

Is my kid faceblind?

Signs of prosopagnosia in children may include:

- Looking "lost" on the playground or at the school pickup.
- Seeming spacey or socially incompetent.
- Being unable to follow the plots of TV shows and movies.
- Getting bullied and being unable to identify the perpetrator.
- Having difficulty keeping track of teammates in sports.
- Being very shy among strangers but outgoing among well-known people.

- Displaying excessive clinginess.
- Tending to get separated from the group (e.g., on school field trips).
- Never using proper names or making introductions.
- Ignoring familiar people when they are out of context (e.g., a teacher at the grocery store).

Should I seek out a formal prosopagnosia diagnosis for myself or my child?

The short answer is no.

Here's the long answer: In the ICD (the International Classifications of Diseases), prosopagnosia barely merits a mention. It's lumped in with all the other visual agnosias, congenital and acquired. That, plus the fact that prosopagnosia is completely absent from the *DSM*, makes the condition basically invisible to the medical community. In the US, you're unlikely to get health insurance reimbursement for diagnosis or treatment. And unless your child is struggling academically, they will probably not be considered eligible for special education or accommodation under the Individuals with Disabilities Education Act (IDEA). (However, perhaps they should be—and if you want to lawyer up and make that case, go for it.)

If you have the money and time to spare, there's no harm in seeking out a diagnosis from a well-informed neurologist, but all this will get you is self-knowledge. A cheaper option is to take a reputable test—like the Cambridge Face Memory Test or the Famous Faces Test—some of which are available for free online, while others are accessible only by volunteering for a study. If you're two deviations below the norm (which is to say, in the bottom 2.5 percent), you, my friend, are faceblind. If you score higher than that but still suspect that you're faceblind, take some more tests, or trust your gut. There's more than one way to be faceblind, and some people are just good at defeating tests. (Visit my website, SadieD.com, for links to tests and studies.)

How can I make life easier for my faceblind child?

First of all, congratulate yourself. You are a very perceptive parent. Then gird yourself, because you are in uncharted territory. There are very few studies on faceblind children, and research on how to help them is just beginning to gear up. The studies that have been done, however, show that faceblindness can be very distressing for kids who have it, and if left to their own devices, faceblind kids are likely to misattribute their social struggles and grow up believing they are unlikable, awkward, or dumb.

Even if your faceblind kid seems well-adjusted, they could probably use your help. Here are a few ideas I have gathered from disability educators, researchers, and my fellow prosopagnosiacs. Please note that these are merely suggestions. Pick and choose the ones that seem appropriate for your particular child and situation.

- Tell your child that they are faceblind in a matter-of-fact, developmentally appropriate way. You might want to compare it to other, more common disorders—for instance, by explaining that prosopagnosia is like dyslexia, but with faces instead of words.
- Use people's names as often as possible. For instance, by saying, "Here comes Jennie. She was in your tap dance class."

- Set up small group playdates. Your child is likely to be more comfortable playing with just one or two kids whose names they can keep track of.
- Teach your child to help themselves—for instance, by asking people their names, and explaining to their friends that they might not recognize them.
- Brainstorm and practice other ways of identifying people, such as by their shape and size, gait, or voice. (But be careful not to insult anyone, as people are often sensitive about their most distinguishing features.)
- Educate their teachers and administrators about prosopagnosia. If your child is okay with it, ask the teacher to explain your child's disorder to the class.
- Ask teachers to use assigned seating and to say students' names as often as possible.
- Consider asking your child's best friend to consistently wear a distinctive accessory.
- Avoid sending your kid to a school with uniforms.

ACKNOWLEDGMENTS

Steve says that I've written a version of him that sounds like a robot, but I had no choice—a more realistic portrayal of my sweet, humble genius of a husband would come off as bragging. This book would never have been possible without his unwavering support and encouragement.

I'd like to thank my family, especially my parents and stepparents, all of whom did a great job—obviously. I am so grateful to my cheerleading grandparents: my grandma Adele, who has read every dang thing I have ever written; my grandma Essie, another lifelong source of unconditional love and support; my grandpa Simon, who taught me to love nature; and my grandpa Mel, who taught me to love books. My brother and sister-in-law have been an unfailing source of thoughtful advice and novel ideas. My insightful nephews have helped me understand what it's like to be able to visualize, see in 3D, and recognize people on sight. I'd also like to thank my baby niece, who, to be honest, wasn't all that helpful, but she is awfully cute.

I owe a great debt of gratitude to the brilliant scientists who patiently answered my many questions and let me know when I got something wrong, such as their hair color. They include Joe DeGutis, Brad Duchaine, Brian Levine, Dennis Levi, Sue Barry,

ACKNOWLEDGMENTS

Ingo Kennerknecht, Doris Tsao, Shahin Nasr, Margaret Livingstone, Nancy Kanwisher, Rodrigo Quian Quiroga, Endel Tulving, Thomas Papathomas, Christopher Tyler, Josh Davis, David Cook, Bruno Rossion, Andrew Oswald, Jian Ding, Rachel Bennetts, Sherryse Corrow, Pete Thompson, Mariska Kret, Barry Sandrew, Adam Zeman, Russell Hurlburt, Alek Krumm, Alexei Dawes, Wilma Bainbridge, Meike Ramon, William von Hippel, Timothy Welsh, H. Moriah Sokolowski, Mary T. Morse, Emma Megla, Alice Lee, Anna Stumps, Maruti Mishra, Hilary Lu, Antônio Mello, and Regan Fry. Thanks also to narrative medicine scholar Danielle R. Spencer and science librarian Tony Stankus. I'm exceedingly grateful to all my Smith College professors, especially Bill Peterson and David Palmer. Devon Price, Lyric Rivera, and Nae: I'm deeply grateful for your help explaining the neurodiversity movement to me. ~~Feel free to blame them for all my mistakes.~~ Any errors are solely my fault.

This project would never have gotten off the ground without the help of my gimlet-eyed agent, Dara Kaye, for baking my half-baked ideas. Thanks also to my genius editors, Tracy Behar and (pinch hitter) Talia Krohn, bottomless wells of wisdom and insight. I owe a debt of gratitude to my former editors, Ian Straus and David Rowell, who ask the hard questions and aren't scared off by a few tears.

Thank you to all my friends who helped me remember our shared experiences. You include Tracy, Joanne, Ann, Anne, Sybil, and Miriam. Special thanks to Sieren, Lea, Holly, and Tom for being excellent readers and providing unflagging moral support; Pam and her partner (David?) for hosting me in Boston; and Heather and Lynia for hosting me in Oakland. I'm also indebted to my fellow journalists who have shared otherwise impossible-to-find information with me: John Beder, Katie

ACKNOWLEDGMENTS

DeRoche, Martine Powers, and Ted Muldoon. A thousand thanks to my "No BS" crew, including Lilian, Angela, Sandy, Dale, Kathy, Cathy, Veronica, Linda, Peace, Shamaine, Alison, Pat, Pam, Jean, Elizabeth, and Mary.

I'm always amazed and deeply appreciative when famous and/or busy people take the time to chat with me. For this book, that includes Sen. John Hickenlooper, Steve Wozniak, Craig Venter, Paul Foot, Andy Pope, Mike Neville, James Blaha, Endri Angjeli, Glenn Alperin, and Larry Kenney.

NOTES

Introduction

1. L. Kay, R. Keogh, T. Andrillon, and J. Pearson, "The Pupillary Light Response as a Physiological Index of Aphantasia, Sensory, and Phenomenological Imagery Strength," *eLife* 11 (2022): e72484, https://doi.org/10.7554/eLife.72484.
2. J. Fulford, F. Milton, D. Salas, A. Smith, A. Simler, C. Winlove, and A. Zeman, "The Neural Correlates of Visual Imagery Vividness—An fMRI Study and Literature Review," *Cortex* 105 (2018): 26–40, https://doi.org/10.1016/j.cortex.2017.09.014.
3. M. Wicken, R. Keogh, and J. Pearson, "The Critical Role of Mental Imagery in Human Emotion: Insights from Fear-Based Imagery and Aphantasia," *Proceedings of the Royal Society B: Biological Sciences* 288, no. 1946 (2021): 20210267, https://doi.org/10.1098/rspb.2021.0267.

1. Grocery Store Epiphany

1. Oliver Sacks, "Face-Blind," *The New Yorker*, August 23, 2010, https://www.newyorker.com/magazine/2010/08/30/face-blind.

2. Three Moms and a Nazi

1. H. D. Ellis and M. Florence, "Bodamer's (1947) Paper on Prosopagnosia," *Cognitive Neuropsychology* 7, no. 2 (1990): 81–105, https://doi.org/10.1080/02643299008253437.
2. I. W. R. Bushnell, "Mother's Face Recognition in Newborn Infants: Learning and Memory," *Infant and Child Development* 10, nos. 1–2 (2001): 67–74, https://doi.org/10.1002/icd.248.
3. J. E. McNeil and E. K. Warrington, "Prosopagnosia: A Face-Specific Disorder," *Quarterly Journal of Experimental Psychology Section A* 46, no. (1) (1993): 1–10, https://doi.org/10.1080/14640749308401064.
4. M. Moscovitch, G. Winocur, and M. Behrmann, "What Is Special About Face Recognition? Nineteen Experiments on a Person with Visual Object Agnosia and Dyslexia but Normal Face Recognition," *Journal of Cognitive Neuroscience* 9, no. 5 (1997): 555–604, https://doi.org/10.1162/jocn.1997.9.5.555.

5. Bill Choisser, "Face Blind!: Chapter 1," last modified November 11, 2014, http://www.choisser.com/faceblind/.
6. Bill Choisser, "Recognizing faces," Google Conversations, May 12, 1996, 12:00:00 a.m., https://groups.google.com/g/alt.support.learning-disab/c/5k2SUd0Zs4k/m/_5GhR_Ez8UcJ.
7. Bill Choisser, "Face Blind!: Appendix A," last modified January 1, 2022, http://www.choisser.com/faceblind/research.html.
8. I. Kennerknecht, T. Grueter, B. Welling, S. Wentzek, J. Horst, S. Edwards, and M. Grueter, "First Report of Prevalence of Non-Syndromic Hereditary Prosopagnosia (HPA)," *American Journal of Medical Genetics* 140A, no. 15 (2006): 1617–22, https://doi.org/10.1002/ajmg.a.31343.

3. Missed Connections

1. A. W. Young, D. C. Hay, and A. W. Ellis, "The Faces That Launched a Thousand Slips: Everyday Difficulties and Errors in Recognizing People," *British Journal of Psychology* 76, no. 4 (1985): 495–523, https://doi.org/10.1111/j.2044-8295.1985.tb01972.x.
2. I. Minio-Paluello, G. Porciello, A. Pascual-Leone, and S. Baron-Cohen, "Face Individual Identity Recognition: A Potential Endophenotype in Autism," *Molecular Autism* 11, no. 1 (2020): 81, https://doi.org/10.1186/s13229-020-00371-0.
3. R. J. Bennetts, N. J. Gregory, J. Tree, C. D. Luft, M. J. Banissy, E. Murray, T. Penton, and S. Bate, "Face Specific Inversion Effects Provide Evidence for Two Subtypes of Developmental Prosopagnosia," *Neuropsychologia* 174 (2022): 108332, https://doi.org/10.1016/j.neuropsychologia.2022.108332.
4. P. Thompson, "Margaret Thatcher: A New Illusion," *Perception* 9, no. 4 (1980): 483–84, https://doi.org/10.1068/p090483.
5. N. L. Segal, A. T. Goetz, and A. C. Maldonado, "Preferences for Visible White Sclera in Adults, Children, and Autism Spectrum Disorder Children: Implications of the Cooperative Eye Hypothesis," *Evolution and Human Behavior* 37, no. 1 (2016): 35–39, https://doi.org/10.1016/j.evolhumbehav.2015.06.006.
6. M. J. Sheehan and M. W. Nachman, "Morphological and Population Genomic Evidence That Human Faces Have Evolved to Signal Individual Identity," *Nature Communications* 5, no. 1 (2014): 4800, https://doi.org/10.1038/ncomms5800.
7. Sheehan and Nachman, "Morphological and Population Genomic Evidence," https://doi.org/10.1038/ncomms5800.
8. T. H. Thelen, "Minority Type Human Mate Preference," *Social Biology* 30, no. 2 (1983): 162–80, https://doi.org/10.1080/19485565.1983.9988531.
9. Z. J. Janif, R. C. Brooks, and B. J. Dixson, "Negative Frequency-Dependent Preferences and Variation in Male Facial Hair," *Biology Letters* 10, no. 4 (2014): 20130958, https://doi.org/10.1098/rsbl.2013.0958.
10. M. E. Kret and M. Tomonaga, "Getting to the Bottom of Face Processing: Species-Specific Inversion Effects for Faces and Behinds in Humans and

NOTES

Chimpanzees (*Pan Troglodytes*)," *PLoS One* 11, no. 11 (2016): e0165357, https://doi.org/10.1371/journal.pone.0165357.

11. V. M. Reid, K. Dunn, R. J. Young, J. Amu, T. Donovan, and N. Reissland, "The Human Fetus Preferentially Engages with Face-Like Visual Stimuli," *Current Biology* 27, no. 12 (2017): 1825–28.e3, https://doi.org/10.1016/j.cub.2017.05.044.
12. O. Pascalis, M. de Haan, and C. A. Nelson, "Is Face Processing Species-Specific During the First Year of Life?" *Science* 296, no. 5571 (2002): 1321–23, https://doi.org/10.1126/science.1070223.
13. D. J. Kelly, P. C. Quinn, A. M. Slater, K. Lee, A. Gibson, M. Smith, L. Ge, and O. Pascalis, "Three-Month-Olds, but Not Newborns, Prefer Own-Race Faces," *Developmental Science* 8, no. 6 (2005): F31–36, https://doi.org/10.1111/j.1467-7687.2005.0434a.x.
14. G. Anzures, A. Wheeler, P. C. Quinn, O. Pascalis, A. M. Slater, M. Heron-Delaney, J. W. Tanaka, and K. Lee, "Brief Daily Exposures to Asian Females Reverses Perceptual Narrowing for Asian Faces in Caucasian Infants," *Journal of Experimental Child Psychology* 112, no. 4 (2012): 484–95, https://doi.org/10.1016/j.jecp.2012.04.005.
15. J. DeGutis, B. Yosef, E. A. Lee, E. Saad, J. Arizpe, J. S. Song, J. Wilmer, L. Germine, and M. Esterman, "The Rise and Fall of Face Recognition Awareness Across the Life Span," *Journal of Experimental Psychology: Human Perception and Performance* 49, no. 1 (2023): 22–33, https://doi.org/10.1037/xhp0001069.
16. J. Liu, J. Li, L. Feng, L. Li, J. Tian, and K. Lee, "Seeing Jesus in Toast: Neural and Behavioral Correlates of Face Pareidolia," *Cortex* 53 (2014): 60–77, https://doi.org/10.1016/j.cortex.2014.01.013.
17. D. Alais, Y. Xu, S. G. Wardle, and J. Taubert, "A Shared Mechanism for Facial Expression in Human Faces and Face Pareidolia," *Proceedings of the Royal Society B: Biological Sciences* 288, no. 1954 (2021): 20210966, https://doi.org/10.1098/rspb.2021.0966.

4. Hacking the System

1. R. Russell, B. Duchaine, and K. Nakayama, "Super-Recognizers: People with Extraordinary Face Recognition Ability," *Psychonomic Bulletin and Review* 16, no. 2 (2009): 252–57, https://doi.org/10.3758/PBR.16.2.252.
2. A. K. Bobak, B. A. Parris, N. J. Gregory, R. J. Bennetts, and S. Bate, "Eye-Movement Strategies in Developmental Prosopagnosia and 'Super' Face Recognition," *Quarterly Journal of Experimental Psychology* 70, no. 2 (2017): 201–17, https://doi.org/10.1080/17470218.2016.1161059.

5. Your Brain's Rosetta Stone

1. R. Quian Quiroga, L. Reddy, G. Kreiman, C. Koch, and I. Fried, "Invariant Visual Representation by Single Neurons in the Human Brain," *Nature* 435, no. 7045 (2005): 1102–7, https://doi.org/10.1038/nature03687.

NOTES

2. M. J. Ison, R. Quian Quiroga, and I. Fried, "Rapid Encoding of New Memories by Individual Neurons in the Human Brain," *Neuron* 87, no. 1 (2015): 220–30, https://doi.org/10.1016/j.neuron.2015.06.016.
3. L. Chang and D. Y. Tsao, "The Code for Facial Identity in the Primate Brain," *Cell* 169, no. 6 (2017): 1013–28.e14, https://doi.org/10.1016/j.cell.2017.05.011.
4. P. Bao, L. She, M. McGill, and D. Y. Tsao, "A Map of Object Space in Primate Inferotemporal Cortex," *Nature* 583, no. 7814 (2020): 103–8, https://doi.org/10.1038/s41586-020-2350-5.
5. C. Zhuang, S. Yan, A. Nayebi, and D. L. K. Yamins, "Unsupervised Neural Network Models of the Ventral Visual Stream," *Proceedings of the National Academy of Sciences* 118, no. 3 (2021): e2014196118, https://doi.org/10.1073/pnas.2014196118.

6. Oblivious Ineptitude

1. M. K. Smith, R. Trivers, and W. von Hippel, "Self-Deception Facilitates Interpersonal Persuasion," *Journal of Economic Psychology* 63 (2017): 93–101, https://doi.org/10.1016/j.joep.2017.02.012.
2. S. C. Murphy, F. K. Barlow, and W. von Hippel, "A Longitudinal Test of Three Theories of Overconfidence," *Social Psychological and Personality Science* 9, no. 3 (2018): 353–63, https://doi.org/10.1177/1948550617699252.
3. S. Bate, A. Adams, and R. J. Bennetts, "Guess Who? Facial Identity Discrimination Training Improves Face Memory in Typically Developing Children," *Journal of Experimental Psychology* 149, no. 5 (2020): 901–13, https://doi.org/10.1037/xge0000689.
4. I. Kennerknecht, N. Pluempe, and B. Welling, "Congenital Prosopagnosia—A Common Hereditary Cognitive Dysfunction in Humans," *Frontiers in Bioscience* 13, no. 8 (2008): 3150–58, https://doi.org/10.2741/2916.

7. Fear, Unmasked

1. R. Le Grand, C. J. Mondloch, D. Maurer, and H. P. Brent, "Early Visual Experience and Face Processing," *Nature* 410, no. 6831 (2001): 890, https://doi.org/10.1038/35073749.
2. R. Le Grand, C. J. Mondloch, D. Maurer, and H. P. Brent, "Expert Face Processing Requires Visual Input to the Right Hemisphere During Infancy," *Nature Neuroscience* 6, no. 10 (2003): 1108–12, https://doi.org/10.1038/nn1121.
3. G. Chatterjee, L. Germine, A. Novick, K. Nakayama, and J. Wilmer, "Poorer Face Recognition in Left-Eye Amblyopes," *Journal of Vision* 12, no. 9 (2012): 484, https://doi.org/10.1167/12.9.484.
4. M. Dombrow and H. M. Engel, "Rates of Strabismus Surgery in the United States: Implications for Manpower Needs in Pediatric Ophthalmology," *Journal of American Association for Pediatric Ophthalmology and Strabismus* 11, no. 4 (2007): 330–35, https://doi.org/10.1016/j.jaapos.2007.05.010.

NOTES

5. E. E. Birch and J. Wang, "Stereoacuity Outcomes Following Treatment of Infantile and Accommodative Esotropia," *Optometry and Vision Science* 86, no. 6 (2009): 647–52, table 1, https://doi.org/10.1097/OPX.0b013e3181a6168d.
6. M. S. Livingstone, R. Lafer-Sousa, and B. R. Conway, "Stereopsis and Artistic Talent: Poor Stereopsis Among Art Students and Established Artists," *Psychological Science* 22, no. 3 (2011): 336–38, https://doi.org/10.1177/0956797610397958.
7. J. A. Bradbury and R. H. Taylor, "Severe Complications of Strabismus Surgery," *Journal of American Association for Pediatric Ophthalmology and Strabismus* 17, no. 1 (2013): 59–63, https://doi.org/10.1016/j.jaapos.2012.10.016.
8. J. M. Baker, C. Drews-Botsch, M. R. Pfeiffer, and A. E. Curry, "Driver Licensing and Motor Vehicle Crash Rates Among Young Adults with Amblyopia and Unilateral Vision Impairment," *Journal of American Association for Pediatric Ophthalmology and Strabismus* 23, no. 4 (2019): 230–32, https://doi.org/10.1016/j.jaapos.2019.01.009.
9. S. E. Kumaran, A. Rakshit, J. R. Hussaindeen, J. Khadka, and K. Pesudovs, "Does Non-Strabismic Amblyopia Affect the Quality of Life of Adults? Findings from a Qualitative Study," *Ophthalmic and Physiological Optics* 41, no. 5 (2021): 996–1006, https://doi.org/10.1111/opo.12864.
10. A. L. Webber, "The Functional Impact of Amblyopia," *Clinical and Experimental Optometry* 101, no. 4 (2018): 443–50, https://doi.org/10.1111/cxo.12663.

8. Student Driver

1. Christopher Tyler, *Bela Julesz 1928–2003: Biographical Memoirs* (Washington, DC: National Academy of Sciences, 2014), https://www.nasonline.org/publications/biographical-memoirs/memoir-pdfs/julesz-bela.pdf.
2. Jeremy Bernstein, *Three Degrees Above Zero: Bell Labs in the Information Age* (Cambridge, UK: Cambridge Univ. Press, 1984), 43.
3. A. B. Zipori, L. Colpa, A. M. F. Wong, S. L. Cushing, and K. A. Gordon, "Postural Stability and Visual Impairment: Assessing Balance in Children with Strabismus and Amblyopia," *PLoS One* 13, no. 10 (2018): e0205857, https://doi.org/10.1371/journal.pone.0205857.
4. S. P. McKee, D. M. Levi, and J. A. Movshon, "The Pattern of Visual Deficits in Amblyopia," *Journal of Vision* 3, no. 5 (2003): 380–405, https://doi.org/10.1167/3.5.5.
5. E. Niechwiej-Szwedo, L. Colpa, and A. M. F. Wong, "Visuomotor Behaviour in Amblyopia: Deficits and Compensatory Adaptations," *Neural Plasticity* 2019 (2019): 6817839, https://doi.org/10.1155/2019/6817839.

9. Sadie Vision

1. Osea Giuntella, Sally McManus, Redzo Mujcic, Andrew J. Oswald, Nattavudh Powdthavee, and Ahmed Tohamy, "The Midlife Crisis" (working paper no. 30442, National Bureau of Economic Research, Cambridge, MA, 2022), https://www.nber.org/papers/w30442.

NOTES

2. A. Weiss, J. E. King, M. Inoue-Murayama, and A. J. Oswald, "Evidence for a 'Midlife Crisis' in Great Apes Consistent with the U-Shape in Human Well-Being," *Proceedings of the National Academy of Sciences* 109, no. 49 (2012): 19949–52, https://doi.org/10.1073/pnas.1212592109.
3. H. Liu, B. Laeng, and N. O. Czajkowski, "Does Stereopsis Improve Face Identification? A Study Using a Virtual Reality Display with Integrated Eye-Tracking and Pupillometry," *Acta Psychologica* 210 (2020): 103142, https://doi.org/10.1016/j.actpsy.2020.103142.

10. Hollywood Meets Science

1. S. Xiao, E. Angjeli, H. C. Wu, E. D. Gaier, S. Gomez, D. A. Travers, G. Binenbaum, R. Langer, D. G. Hunter, M. X. Repka, and Luminopia Pivotal Trial Group, "Randomized Controlled Trial of a Dichoptic Digital Therapeutic for Amblyopia," *Ophthalmology* 129, no. 1 (2022): 77–85, https://doi.org/10.1016/j.ophtha.2021.09.001.
2. O. Uretmen, S. Egrilmez, S. Kose, K. Pamukçu, C. Akkin, and M. Palamar, "Negative Social Bias Against Children with Strabismus," *Acta Ophthalmologica Scandinavica* 81, no. 2 (2003): 138–42, https://doi.org/10.1034/j.1600-0420.2003.00024.x.
3. S. M. Mojon-Azzi, A. Kunz, and D. S. Mojon, "Strabismus and Discrimination in Children: Are Children with Strabismus Invited to Fewer Birthday Parties?" *British Journal of Ophthalmology* 95, no. 4 (2011): 473–76, https://doi.org/10.1136/bjo.2010.185793.
4. E. A. Paysse, E. A. Steele, K. M. B. McCreery, K. R. Wilhelmus, and D. K. Coats, "Age of the Emergence of Negative Attitudes Toward Strabismus," *Journal of American Association for Pediatric Ophthalmology and Strabismus* 5, no. 6 (2001): 361–66, https://doi.org/10.1067/mpa.2001.119243.
5. A. P. Akay, B. Cakaloz, A. T. Berk, and E. Pasa, "Psychosocial Aspects of Mothers of Children with Strabismus," *Journal of American Association for Pediatric Ophthalmology and Strabismus* 9, no. 3 (2005): 268–73, https://doi.org/10.1016/j.jaapos.2005.01.008.
6. S. M. Mojon-Azzi and D. S. Mojon, "Strabismus and Employment: The Opinion of Headhunters," *Acta Ophthalmologica* 87, no. 7 (2009): 784–88, https://doi.org/10.1111/j.1755-3768.2008.01352.x.
7. S. M. Mojon-Azzi, W. Potnik, and D. S. Mojon, "Opinions of Dating Agents About Strabismic Subjects' Ability to Find a Partner," *British Journal of Ophthalmology* 92, no. 6 (2008): 765–69, https://doi.org/10.1136/bjo.2007.128884.
8. A. N. Buffenn, "The Impact of Strabismus on Psychosocial Health and Quality of Life: A Systematic Review," *Survey of Ophthalmology* 66, no. 6 (2021): 1051–64, https://doi.org/10.1016/j.survophthal.2021.03.005.

NOTES

11. Video Game Therapy

1. T. L. Ooi and Z. J. He, "Space Perception of Strabismic Observers in the Real World Environment," *Investigative Ophthalmology and Visual Science* 56, no. 3 (2015): 1761–68, https://doi.org/10.1167/iovs.14-15741.
2. B. Bridgeman, "Restoring Adult Stereopsis: A Vision Researcher's Personal Experience," *Optometry and Vision Science* 91, no. 6 (2014): e135–39, https://doi.org/10.1097/OPX.0000000000000272.
3. Morgen Peck, "How a Movie Changed One Man's Vision Forever," Neuroscience, BBC Future, July 18, 2012, https://www.bbc.com/future/article/20120719-awoken-from-a-2d-world.
4. J. Wirtz, "Creativity in 3D: Poets and Scientists Converge on Writerly Invention," *Interdisciplinary Science Reviews* 39, no. 1 (2014): 62–72, http://dx.doi.org/10.1179/0308018813Z.00000000068.

12. We're All Making the Same Mistake

1. Martin Brookes, *Extreme Measures: The Dark Visions and Bright Ideas of Francis Galton* (New York: Bloomsbury, 2004).
2. D. Burbridge, "Galton's 100: An Exploration of Francis Galton's Imagery Studies," *British Journal for the History of Science* 27, no. 4 (1994): 443–63, http://www.jstor.org/stable/4027625.
3. Karl Pearson, *Life, Letters and Labours of Francis Galton*, vol. 2 (Cambridge, UK: Cambridge Univ. Press, 1924), 194.
4. W. F. Brewer and M. Schommer-Aikins, "Scientists Are Not Deficient in Mental Imagery: Galton Revised," *Review of General Psychology* 10, no. 2 (2006): 130–46, https://doi.org/10.1037/1089-2680.10.2.130.
5. Carl Zimmer, "Picture This? Some Just Can't," *New York Times*, June 22, 2015, https://www.nytimes.com/2015/06/23/science/aphantasia-minds-eye-blind.html.
6. New Scientist, *Your Conscious Mind: Unravelling the Mystery of the Human Brain* (London: John Murray Press, 2017).
7. Carl Zimmer, "The Brain: Look Deep into the Mind's Eye," *Discover Magazine*, March 22, 2010, https://www.discovermagazine.com/mind/the-brain-look-deep-into-the-minds-eye.
8. New Scientist, *Your Conscious Mind*.
9. A. Z. J. Zeman, S. Della Sala, L. A. Torrens, V.-E. Gountouna, D. J. McGonigle, and R. H. Logie, "Loss of Imagery Phenomenology with Intact Visuo-Spatial Task Performance: A Case of 'Blind Imagination,'" *Neuropsychologia* 48, no. 1 (2010): 145–55, https://doi.org/10.1016/j.neuropsychologia.2009.08.024.
10. M. Matsuhashi and M. Hallett, "The Timing of the Conscious Intention to Move," *European Journal of Neuroscience* 28, no. 11 (2008): 2344–51, https://doi.org/10.1111/j.1460-9568.2008.06525.x.

11. P. Johansson, L. Hall, S. Sikström, and A. Olsson, "Failure to Detect Mismatches Between Intention and Outcome in a Simple Decision Task," *Science* 310, no. 5745 (2005): 116–19, https://doi.org/10.1126/science.1111709.
12. J. B. Watson, "Psychology as the Behaviorist Views It," *Psychological Review* 20, no. 2 (1913): 158–77, https://doi.org/10.1037/h0074428.
13. B. Faw, "Conflicting Intuitions May Be Based on Differing Abilities: Evidence from Mental Imaging Research," *Journal of Consciousness Studies* 16, no. 4 (2009): 45–68, https://psycnet.apa.org/record/2009-05537-003.
14. J. B. Watson, "Image and Affection in Behavior," *Journal of Philosophy, Psychology, and Scientific Methods* 10, no. 16 (1913): 423n3, https://www.jstor.org/stable/2012899?seq=4.
15. Watson, "Image and Affection in Behavior," 424.

13. Quantifying Quirkiness

1. William James, "The Stream of Consciousness," in *Psychology* (Cleveland, OH: World, 1948), ch. 11.
2. C. L. Heavey and R. T. Hurlburt, "The Phenomena of Inner Experience," *Consciousness and Cognition* 17, no. 3 (2008): 798–810, https://doi.org/10.1016/j.concog.2007.12.006.
3. Stephanie Doucette and Russell T. Hurlburt, "Inner Experience in Bulimia," in *Sampling Inner Experience in Disturbed Affect: Emotions, Personality, and Psychotherapy* (New York: Springer, 1993), https://doi.org/10.1007/978-1-4899-1222-0_10.
4. J. Craig Venter, *A Life Decoded* (New York: Viking Penguin, 2007), 27, 14.

14. Triangulating the Truth

1. M. Wicken, R. Keogh, and J. Pearson, "The Critical Role of Mental Imagery in Human Emotion: Insights from Fear-Based Imagery and Aphantasia," *Proceedings of the Royal Society B: Biological Sciences* 288, no. 1946 (2021): 20210267, https://doi.org/10.1098/rspb.2021.0267.
2. B. Laeng and U. Sulutvedt, "The Eye Pupil Adjusts to Imaginary Light," *Psychological Science* 25, no. 1 (2014): 188–97, https://doi.org/10.1177/0956797613503556.
3. W. A. Bainbridge, Z. Pounder, A. F. Eardley, and C. I. Baker, "Quantifying Aphantasia Through Drawing: Those Without Visual Imagery Show Deficits in Object but Not Spatial Memory," *Cortex* 135 (2021): 159–72, https://doi.org/10.1016/j.cortex.2020.11.014.
4. Dan P. McAdams, *The Stories We Live By: Personal Myths and the Making of the Self* (New York: William Morrow, 1993).
5. C. Daniel and O. J. Mason, "Predicting Psychotic-Like Experiences During Sensory Deprivation," *Biomed Research International* 2015 (2015): 439379, https://doi.org/10.1155/2015/439379.

NOTES

15. Gullible's Travails

1. S. J. Babb and J. D. Crystal, "Episodic-Like Memory in the Rat," *Current Biology* 16, no. 13 (2006): 1317–21, https://doi.org/10.1016/j.cub.2006.05.025.
2. D. J. Palombo, C. Alain, H. Söderlund, W. Khuu, and B. Levine, "Severely Deficient Autobiographical Memory (SDAM) in Healthy Adults: A New Mnemonic Syndrome," *Neuropsychologia* 72 (2015): 105–18, https://doi.org/10.1016/j.neuropsychologia.2015.04.012.
3. E. Tulving, "Memory and Consciousness," *Canadian Psychology/Psychologie Canadienne* 26, no. 1 (1985): 1–12, https://www.apa.org/pubs/journals/features/cap-h0080017.pdf.
4. L. Vanaken, P. Bijttebier, R. Fivush, and D. Hermans, "Narrative Coherence Predicts Emotional Well-Being During the COVID-19 Pandemic: A Two-Year Longitudinal Study," *Cognition and Emotion* 36, no. 1 (2022): 70–81, https://doi.org/10.1080/02699931.2021.1902283.
5. L. A. King, C. K. Scollon, C. Ramsey, and T. Williams, "Stories of Life Transition: Subjective Well-Being and Ego Development in Parents of Children with Down Syndrome," *Journal of Research in Personality* 34, no. 4 (2000): 509–36, https://doi.org/10.1006/jrpe.2000.2285.

16. Comic Incompetence Is the Brand

1. R. Siugzdaite, J. Bathelt, J. Holmes, and D. E. Astle, "Transdiagnostic Brain Mapping in Developmental Disorders," *Current Biology* 30, no. 7 (2020): 1245–57.e4, https://doi.org/10.1016/j.cub.2020.01.078.
2. G. J. August and B. D. Garfinkel, "Comorbidity of ADHD and Reading Disability Among Clinic-Referred Children," *Journal of Abnormal Child Psychology* 18, no. 1 (1990): 29–45, https://doi.org/10.1007/BF00919454.
3. C. Hours, C. Recasens, and J.-M. Baleyte, "ASD and ADHD Comorbidity: What Are We Talking About?" *Frontiers in Psychiatry* 13 (2022): 837424, https://doi.org/10.3389/fpsyt.2022.837424.
4. I. Minio-Paluello, G. Porciello, A. Pascual-Leone, and S. Baron-Cohen, "Face Individual Identity Recognition: A Potential Endophenotype in Autism," *Molecular Autism* 11, no. 1 (2020): 81, https://doi.org/10.1186/s13229-020-00371-0.
5. M. Hartston, G. Avidan, Y. Pertzov, and B.-S. Hadad, "Weaker Face Recognition in Adults with Autism Arises from Perceptually Based Alterations," *Autism Research* 16, no. 4 (2023): 723–33, https://doi.org/10.1002/aur.2893.

INDEX

Note: Italic page numbers refer to illustrations.

abstract concepts
  severely deficient autobiographical memory and, 238
  visual experiences related to, 134
  visualization skills and, 180, 185
Adams, Douglas, 97–99
ADHD
  activist movements and, 90
  prevalence of, 6
  studies of, 251, 252
  visual imagination and, 186
agnosia. *See also* faceblindness
  meaning of, 253
  topographical agnosia, 44, 168, 252
  visual agnosia, 30, 31, 268
ahegao (cross-eyed "O" face), 155
Alzheimer's disease, 244
amblyopia (lazy eye)
  baby's age and, 158, 158n
  causes of, 140
  as cerebral vision impairment, 163–64
  color vision and, 141–42
  discerning details at distance and, 142
  distorted sense of space and, 169
  driving and, 126–27, 143
  eye patches for, 145, 148
  left-eye amblyopia, 103, 126–27, 176
  prevalence of, 147
  strabismus and, 103
  treatments for, 137–38, 142–43, 146–47, 148
  vision therapy for, 161–62
  VR (virtual-reality) technology for, 148–49
American Psychological Association, 17, 17n, 70, 104
Angjeli, Endri, 149
animals
  behaviorism and, 190, 191–93, 191n
  chimpanzee studies, 51–52, 53, 135, 235
  imagining subjective experiences of, 214–15
  macaque studies, 71–72, 76, 235
  memory and, 235–36
Aniston, Jennifer, 70–71
aphantasia (blind mind's eye)
  aural imagination and, 187
  autobiographical memories and, 186
  Wilma Bainbridge's studies on, 220–23, *221*
  binocular rivalry and, 217
  brain differences and, 259, 263–64
  faceblindness associated with, 187
  memory and, 186, 208–9
  online debates and, 207–8
  Joel Pearson's studies of, 216–17, 241–42
  sensory deprivation and, 226–28

INDEX

aphantasia *(cont.)*
  severely deficient autobiographical memory (SDAM) and, 238, 239–41, 240n, 258
  studies of, 218–19
  subgroups of aphants, 187
  Craig Venter and, 208–11
  visual imagination and, 7, 176, 182–83, 186, 187, 200, 201, 208–9, 211, 216–17, 220–23, 227, 258, 263
apperceptive prosopagnosia, 45
applied behavioral analysis, 191–92
Aristotle, 194
artists, and stereoblindness, 11–12, 104–5, 128
associative prosopagnosia, 45
Athinoula A. Martinos Center, 132
auditory processing disorders, 44, 252
aural imagination, 187, 194
autism, 45, 90, 251, 252, 256
autosomal dominant disorders, 95
autostereogram, 122
Avatar (film), 145
Avatar 2 (3D film), 173–74, 209

Baccei, Tom, 122
Bainbridge, Wilma, 218, 220–24, *221*, 258, 262–64
Barry, Susan R.
  on stereoblindness, 108–9, 134, 174–75
  stereovision of, 109, 125, 134, 136, 142, 168–69, 174, 175–76
  vision therapy of, 108–9, 125, 168–69, 172
behavioral optometry centers, 137
behaviorism, 190, 191–94, 191n
Bell Labs, 119–20
Ben Franklin Effect, 192
Bennetts, Rachel, 92–94
Benton Facial Recognition Test, 43
Berry, Halle, 70
Bezos, Jeff, 21

big-picture thinking, 211, 238
binocular disparity (3D vision), 141
binocular rivalry, 217
birdcalls, 175
bird-watching, 169–70, 172–73
Blaha, James, 145–49, 167
Bodamer, Joachim, 26–27, 26n, 29–30, 73
brain. *See also* fusiform face area (FFA)
  aphantasia and, 259, 263–64
  complexity of, 253
  daisy-chain network structure and, 251
  excess neuroplasticity and, 136
  hemispheres of, 36, 102–3, 102n
  high-resolution images of, 126
  hippocampi of, 71, 233, 234, 237–38, 263, 264
  hub-and-spoke network structure and, 251
  medial temporal lobe, 263
  MRI machines and, 131–33
  object recognition and, 53–54, 72–73
  visual-processing streams and, 219–20, *219*
brain injury, in acquired prosopagnosia, 45
Breaker experiment, 146–47
Breakfast Table Questionnaire, 179–80
Bridgeman, Bruce, 172–73
Bryant, Kobe, 70
bulimia, 199
butts, identification of, 50–52, 53

Caballero, Alexander, 63–64
Cambridge Face Memory Test (CFMT), 17–20, 42–43, 45, 63, 102, 268
Cambridge Face Perception Test (CFPT), 43, *43*, 45
cerebral vision impairment (CVI), 163–65

INDEX

Chagall, Marc, 11
Chang, Le, 71–72
chimpanzees
  butt identification and, 51–52, 53
  happiness curve and, 135
  memory and, 235
Choisser, Bill, 31–33, 35, 44
Choisser, Edna, 31–32
Clinton, Hillary, 20
Close, Chuck, 45
Cochrane, Kent, 233–36, 243, 244
cognitive disorders, 6
colorblindness, 6, 26
color vision, amblyopia associated with deficiency in, 141–42
computer face-recognition systems, 36
concepts. *See also* abstract concepts
  neurons linking associated concepts, 71
conscious experience, diversity of, 180
consciousness
  forms of, 195
  multiple modes of, 199
  types of, 198
Corrow, Sherryse, 92–93
COVID-19 pandemic, 101

darkness, and stereoblindness, 175
Darwin, Charles, 177–78, 180
Davis, Josh, 62–64
Dawes, Alexei, 217–18
definitive statements, 253n
DeGutis, Joe
  faceblindness research of, 20, 22, 42, 49, 53–55, 57, 75–77, 133, 162
  face-recognition training program and, 58–60, 66–67
  Vision Sciences Society and, 260
depth perception
  stereoblindness and, 12, 103, 106–8, 118, 128, 166
  triangulation and, 120–21, 124

Descriptive Experience Sampling (DES), 198–99, 201, 203, 204–7
development, critical periods of, 125
developmental amnesiacs, 237, 237n
developmental prosopagnosia (DP), 35, 45–46, 95, 254
Diana, Princess, 171
Ding, Jian, 161–63, 170–72
Dingfelder, Katharine, 186
Dingfelder, Lynn (stepmother), 156, 159, 240
Dingfelder, Sadie
  aphantasia of, 176, 200–201, 207–8, 218–19, 252, 258, 263, 265
  appearance of, 155–56, 200–201, 223
  being lost and, 43, 44
  birth trauma and, 104, 158
  car-buying experiences of, 113–16, 151–52
  COVID-19 pandemic and, 101
  decision to move to the country, 110–11, 113–14
  driver-assist features and, 151–52, 153
  driver's ed experience of, 97–100
  driving fears of, 7, 109, 110
  driving lessons of, 110–11, 113, 116, 117–18, 126–29, 140, 149–52
  driving safety and, 143
  eye-realignment surgeries of, 104, 107, 136, 140, 153
  faceblindness diagnosis and, 53, 75–80, 82–83
  faceblindness of, 3–5, 6, 7, 13–16, 28–29, 36–38, 40, 53, 55, 57, 66, 201, 207, 214, 252, 253, 254, 260, 265
  faceblindness research of, 16–21
  faceblindness story of, 22, 75, 89, 100–101
  face-recognition training program and, 58–61, 65–67, 76
  father's guilt over falls of, 95, 157–59

INDEX

Dingfelder, Sadie *(cont.)*
  father's reaction to eye surgeries of, 107
  father's reaction to faceblindness of, 6–7, 37–38, 77–78, 81, 83
  father's reaction to hiking of, 168
  father's reaction to stereoblindness of, 106–7
  father's relationship with, 141, 143, 156–59, 259–62
  finding loopholes and, 261
  fMRI of, 53, 55–56
  freelance work of, 101, 135
  friendships of, 56, 90, 138–40, 155, 170, 180–82, 184–86, 188–92, 214, 226, 241
  genealogy research of, 101–2
  Harvard study on faceblindness and, 22–23, 42–43, 53, 55–56
  hiking experiences of, 168
  husband and grocery story incident, 13–14
  husband as driver for, 3
  husband on face-recognition training, 59, 65–66
  husband on vision therapy, 167
  husband's attitude toward seeing old friends, 56
  husband's bird-watching, 169–70
  husband's cooking skills, 157
  husband's dishwashing and, 214
  husband's driving lessons for, 110–11, 127–29, 149–50
  husband's face memory test and, 18–20, 21
  husband's reaction to faceblind diagnosis, 78, 80
  husband's visual imagination, 185
  husband's wedding location discussion and, 44
  left-eye amblyopia of, 103, 126–27, 176
  life hacks of, 42, 138–39
  loneliness of, 7, 75, 80, 81–82, 87, 201
  LSD experience of, 229–31
  lying still ability of, 55–56, 132, 137, 203n
  memory deficits of, 40–42, 49, 55, 139–40, 226
  middle age theory of, 135–36, 137
  middle school experiences of, 78–81
  migraines of, 48, 162, 170
  mind-blindness of, 207
  neurodiversity of, 7–8, 11–12, 264–65
  note-taking habit of, 224–25, 244, 265
  ophthalmologist appointments of, 105–6, 107
  parking car and, 153
  rental car experience of, 247–49
  sense of humor and, 88
  7-Tesla MRI machine and, 131–33
  severely deficient autobiographical memory (SDAM) diagnosis, 264–65
  stereoacuity of, 166, 172
  stereoblindness of, 7, 11–12, 103, 105, 106, 107–10, 118, 122, 134, 157, 161–63, 166, 168, 169, 201, 252, 265
  storytelling of, 225–26, 244–46, 261, 265
  strabismus of, 155, 158
  as student, 202–3
  tolerance of ambiguity, 44, 265
  vision therapy of, 165–67, 170–72
  visual imagination of, 184–85
  writing ability of, 203
Dingfelder, Saul (brother), 20, 157, 159, 182, 186–87
disability rights, 163–65
dogs, and sense of smell, 215
Down syndrome, 245
DSM (Diagnostic and Statistical Manual of Mental Disorders), 252, 253, 268

INDEX

Duchaine, Brad, 33–36, 95, 254, 256–57
Duchaine, Dick, 33–34
Duchaine, Pam, 33–34
dyslexia, 6, 251, 252, 269

eating disorders, and inner lives study, 199–200
emotions
  memory and, 244
  visualization linked with, 217, 242
episodic memory, 234, 235–37, 238, 243, 244, 245
esotropia (cross-eye), 155
eugenics, 178
exotropia (outward-pointing eyes, walleye), 128, 154, 155
eye-realignment surgery, 104–5, 107, 136, 140, 153
eye-tracking studies, 64–65

faceblindness
  accommodations for, 91–92
  acquired prosopagnosia and, 35, 45, 87, 95
  autism associated with, 45, 252
  Joachim Bodamer on, 26–30
  as cerebral vision impairment, 163–64
  children with, 92–95, 267–70
  Joe DeGutis's research on, 20, 22, 42, 49, 53–55, 57, 75–77, 133, 162
  developmental prosopagnosia and, 35, 45–46, 95, 254
  diagnosis of, 6, 93, 268–69
  disclosure of, 90
  dyslexia associated with, 45
  environmental studies on, 102
  experience of, 216, 267–68
  Facebook group on, 57–58
  fusiform face area and, 5, 30, 31, 73
  genetic component of, 95
  Harvard study on, 22–23, 42–43, 53, 55–56
  hemi-prosopometamorphopsia, 36
  humor and, 88
  left-eye amblyopia and, 103
  object recognition and, 53–54
  prevalence of, 5–6
  public awareness of, 90
  recognition of mother and, 28–29, 31–32
  Oliver Sacks on, 16
  signs of prosopagnosia in children, 267–68
  studies of, 16–21, 30–36, 53, 256, 269
  symptoms of, 5, 5n
  tips for parents of faceblind child, 269–70
  topographical agnosia and, 44
  TV and film plots and, 32, 173–74, 215, 240, 267
  W. J. case, 30–31
Facebook, faceblindness support group of, 90
face-identification software, 91
face-inversion effect, 46, *46*, 47–48, *47*, 64
face pareidolia, 52–53
face perception
  improvement of, 76–77, 94, 137, 254
  patterns of distortion and, 255–56
  stereovision improvement and, 137
  studies of, 254–56
face-recognition training program, 58–61, *59*, 65–67, 76
facial recognition
  animal-face recognition and, 30–31
  Cambridge Face Memory Test and, 17–20, 42–43, 45, 63, 102
  computer face-recognition systems, 36
  face-identification skills and, 52–53, 94, 137
  face processing and, 92
  failures of, 252
  forgetting names compared to, 4, 5, 40–41

INDEX

facial recognition *(cont.)*
   holistic perception and, 46, 51, 61, 64, 102
   human-face perception, 31
   macaques study and, 71–72, 76
   other-race effect, 52
   social interactions and, 84, 269–70
   super recognizers and, 19, 62–65, 89
Famous Faces Test, 20, 268
fat-shaming, 200
Faw, Bill, 193–94
FDA, and digital therapeutics, 147, 148, 149
FFA. *See* fusiform face area (FFA)
50 First Dates (film), 243
Finding Dory (film), 243
fission–fusion societies, 50
fixed-ratio reinforcement paradigm, 193
Float Sixty, 226–27
fMRI machines
   aphantasia studies and, 218–19
   brain and, 131–33
   faceblindness study and, 53, 55–56
   visualization analysis and, 10
Foot, Paul, 88
free will, 189
full-face prosopometamorphopsia (PMO), 254–56, *255*
fusiform face area (FFA)
   acquired prosopagnosia and, 27, 95
   evolution of, 50
   faceblindness and, 5, 30, 31, 73
   faces activating, 263
   MRIs of, 132
   phantom faces and, 53
   right hemisphere and, 36, 54, 95, 103
   thickness of, 76, 78
   topographical agnosia and, 44
   visual imagination and, 184
   visual input from left eye and, 103
   visual-processing streams and, 220

Galton, Frances, 177–80, 188, 195
Gellar, Sarah Michelle, 70
grandmother neuron hypothesis, 70
Grant, Hugh, 30
Guess Who? game, 94

H&W, 123–26, *124*, 132, 172
happiness curve, U-shaped trajectory of, 135–36, 135n
Hartston, Marissa, 256
hearing, 8, 175, 187, 198
Hemingway, Ernest, 209
Hemi-PMO (hemi-prosopometamorphopsia), 36, 254
Hickenlooper, John, 85–88
highly superior autobiographical memory (HSAM), 238–39
hippocampi, 71, 233, 234, 237–38, 263, 264
Hitler, Adolf, 27
hominids, homogenous faces of, 50
Hopper, Edward, 11
Hubel, David H., 123–26, 132, 172
Hugo (3D film), 172
human consciousness, 195, 198, 199
human faces, animal faces compared to, 49–50
human genome, 208
human visual system, mapping of, 141
Hungary, 118–19
Hurlburt, Russell, 197–99, 201–2, 204–7
hyperphantasia, 182

ICD (International Classifications of Diseases), 268
image streaming, 227
imagination, characteristics of, 9
Individuals with Disabilities Education Act (IDEA), 268
inner hearing, 8, 175, 198

INDEX

inner life
   studies of, 197–99, 202–7
   variations in, 8–9
inner monologues, 8, 199
inner seeing, 198, 199
inner speech, 194
inner visualizations, 8–9, 199–200, 201
internal experiences, study of, 10–11
International Classifications of Diseases (ICD), 268
introspection, 197
Izzard, Eddie, 88

Jamaica Plain VA Medical Center, Boston, 42–43, 48
James, William, 197
Jennifer Aniston neuron, 70–71
Jobs, Steve, 87
Johansson, Scarlett, 20
Julesz, Bela, 119–24
Justice, Jim, 114

Kennerknecht, Ingo, 26n, 35, 95
Klimt, Gustav, 11
Kret, Mariska, 50–51, 53
Krumm, Alek, 204–7
Kudrow, Lisa, 71

labels, of neurodiverse people, 253, 254, 259
lazy eye. *See* amblyopia (lazy eye)
Leclerc, George-Louis, Comte de Buffon, 138
Lee, Alice, 20, 42–44, 53, 58, 60
left-handedness, 91
left-right confusion, 252
lethonomia (forgetting names), 4, 5, 40–42, 55, 89, 91
Levi, Dennis, 137–38
Levi Eye Lab, 161–62, 164, 166, 170
Levine, Brian, 236–44, 250, 258, 264–65
Li, Elizabeth, 255
Lion King, The (film), 154

Livingstone, Margaret, 104, 128
Lowes, Judith, 94–95
Lu, Hilary, 165–66, 171–72
Luminopia, 148–49

McAdams, Dan, 226, 245
macaques
   facial recognition study and, 71–72, 76
   memory and, 235
McConachie, H. R., 34n
McKinnon, Susie, 236–37
McMaster University, 102
Magic Eye puzzles, 122–23
Man Ray, 11
Marcos, Imelda, 31
Markle, Meghan, 49
Marshall, John, 179–80, 188
Megla, Emma, 222–23, 262–63
Mello, Antônio, 254–56
Memento (film), 240, 243
memory
   animals and, 235–36
   aphantasia and, 186, 208–9
   characteristics of, 9
   Kent Cochrane and, 233–36
   episodic memory, 234, 235–37, 238, 243, 244, 245
   function of, 244
   highly superior autobiographical memory (HSAM), 238–39
   rote memorization, 210–11, 239
   semantic memory, 234, 238
   severely deficient autobiographical memory (SDAM), 7, 237–44, 240n, 252, 258–59, 264–65
   super recognizers and, 65
   visual memory, 208–9, 214, 220–22, *221*
   Steve Wozniak's study of, 87
mental rotation test, 183–84, *183*
mental time travel, subjective experience of, 236
Meta Quest, 165

INDEX

mind-blindness, 207
mind's ear, hearing in, 8, 187
monkeys, visual system research on, 141
Murphy, Cillian, 201
Musk, Elon, 6
MX, case of, 182–84
myelinization, 76

Nagel, Thomas, 214
names, remembering of, 4, 5, 40–42, 55, 89, 91
Nasr, Shahin, 131–33, 140–43, 158, 172, 176
National Institutes of Health, 208
neural pruning, 76, 78
neurodevelopmental disorders, 44, 252
neurodiverse people
    accommodation of, 259
    activist movements and, 90, 163–65, 253
    brain structure of, 251, 264
    cognitive profiles of, 252
    experience of, 215–16, 246
    finding loopholes and, 261
    labels and, 253, 254, 259
    prevalence of, 8–9, 10, 214
    studies of, 250–52, 251n
neuroplasticity, 125, 136
neurotypical people
    asymmetry of hippocampi and, 237
    binocular rivalry and, 217
    brain structure of, 251
    episodic and semantic memory and, 238
    face-inversion effect and, 46
    studies of, 250–52, 251n
    super recognizers compared to, 64
    visual imagination and, 220
    visualization in reading and, 9, 218
Neville, Mike, 62–64
New Scotland Yard Super Recogniser Unit, 63

Obama, Barack, 224
object recognition, 53–54, 72–73, 120–21
occipital lobe, 45
occipital place areas (OPAs), 263
ocular dominance columns, 125, 132
Oculus Rift, 146, 165
OPAs (occipital place areas), 263
orangutans, and happiness curve, 135
Oswald, Andrew, 135, 135n
other-race effect, 52, 256

parahippocampal place, and topographical agnosia, 44
Pearson, Joel, 216–17, 241–42
people with disabilities
    education of, 268
    Nazi government's murder of, 26n
    rights of, 163–65
peripheral vision, 128
Pitt, Brad, 20, 70–71, 90
Plato, 27
PMO (prosopometamorphopsia), 254–56, *255*
Pope, Andy, 61–62, 65
Powers, Martine, 75
prefrontal cortex, 184, 227
premotor cortex, 189
primary visual cortex (V1)
    research on, 141, 219
    retinotopic organization of, 124–25, *124*
propioception, 127
ProsoKids Weekend, 93–94
prosopagnosia. *See* faceblindness
prosopometamorphopsia (PMO), 254–56, *255*
psychology
    behaviorism, 190, 191–94, 191n
    and inner experiences studies, 9–10
PsycNet, 17
pupillary dilation response, 217

INDEX

visualization
  emotions linked with, 217, 242
  fMRI analysis of, 10
  learning methods for, 11, 12
  reading and, 8, 9, 218
  visual-processing streams in brain, 219
visual memory
  lack of, 208–9, 214
  research on, 220–22, *221*
visual processing, difficulties in, 175
visual-processing streams
  dorsal stream, 219–20, *219*
  ventral stream, 219, *219*, 220
Vivid Vision, 147–48, 165–67, 170
Von Hippel, William, 83–85
VR (virtual-reality) headsets, for stereoblindness, 138, 146

VR (virtual-reality) technology
  for amblyopia, 148–49
  for stereoblindness, 138, 146–48

Washington Post, 21–22, 75, 89
Washington Post Express, 21, 100–101
Waters, John, 14–15, 17
Watson, John, 190–91, 193–94
Wiesel, Torsten N., 123–26, 132, 172
Winfrey, Oprah, 200
World War II
  H. A. (lieutenant) and, 25–27, 30
  Uffz. S. and, 27, 30
Wozniak, Steve, 87–88
Wyeth, Andrew, 11

Yeltsin, Boris, 31

Zeman, Adam, 182–84, 186–88, 195

About the Author

Sadie Dingfelder is a science journalist who is currently obsessed with hidden neurodiversity and science-based answers to the question: If you were beamed into the mind of another person or animal, what would that be like? She spent six years as a reporter for the *Washington Post Express*, the big *Post*'s daily commuter paper. As an arts reporter, she got to interview many of her heroes, including Yo-Yo Ma, Maria Bamford, Dave Barry, Ann Patchett, and Helen Macdonald. She also penned a biweekly column, The Staycationer, where she wrote about her DC adventures, including having a walk-on part in the Washington Ballet's *Nutcracker*, auditioning to be a "Nationals Racing President," and playing one of the Smithsonian's priceless Stradivarius violins. She regularly contributed features and personal essays to other sections of the *Post*. Her most famous story was about a crane who fell in love with her zookeeper. Also popular was Dingfelder's review of every single bathroom on the National Mall. Her freelance writing has appeared in *National Geographic, National Geographic Traveler, Washingtonian,* and *Washington City Paper*. Prior to working at the *Post*, Dingfelder spent almost a decade as the senior science writer for the American Psychological Association's *Monitor on Psychology* magazine, covering new findings in neuroscience, cognitive science, and ethology. **SadieD.com**